The Mystery of Evolutionary Mechanisms

The Mystery of Evolutionary Mechanisms

Darwinian Biology's Grand Narrative of Triumph
and the Subversion of Religion

Robert F. Shedinger

CASCADE *Books* • Eugene, Oregon

THE MYSTERY OF EVOLUTIONARY MECHANISMS
Darwinian Biology's Grand Narrative of Triumph and the Subversion of Religion

Copyright © 2019 Robert F. Shedinger. All rights reserved. Except for brief quotations in critical publications or reviews, no part of this book may be reproduced in any manner without prior written permission from the publisher. Write: Permissions, Wipf and Stock Publishers, 199 W. 8th Ave., Suite 3, Eugene, OR 97401.

Cascade Books
An Imprint of Wipf and Stock Publishers
199 W. 8th Ave., Suite 3
Eugene, OR 97401

www.wipfandstock.com

PAPERBACK ISBN: 978-1-5326-5833-4
HARDCOVER ISBN: 978-1-5326-5834-1
EBOOK ISBN: 978-1-5326-5835-8

Cataloguing-in-Publication data:

Names: Shedinger, Robert F., author.

Title: The mystery of evolutionary mechanisms : Darwinian biology's grand narrative of triumph and the subversion of religion / Robert F. Shedinger.

Description: Eugene, OR: Cascade Books, 2019. | Includes bibliographical references and index.

Identifiers: ISBN 978-1-5326-5833-4 (paperback). | ISBN 978-1-5326-5834-1 (hardcover). | ISBN 978-1-5326-5835-8 (ebook).

Subjects: LCSH: Religion and Science. | Evolution (Biology). | Natural selection. | Intelligent design (Teleology). | Naturalism—Religious aspects.

Classification: BL240.3 S541 2019 (print). | BL240.3 (ebook).

Manufactured in the U.S.A.   JUNE 20, 2019

To the memory of Bruce Wrightsman, who was always willing to court controversy in the pursuit of truth

Contents

Acknowledgments | ix

Introduction | xi

1. Darwinian Biology's Grand Narrative of Triumph and the Subversion of Religion | 1

2. Reframing Darwin and Darwinism | 32

3. The Eclipse and Recovery of Darwinism | 66

4. Reframing the Modern Synthesis: The Reality of Natural Selection | 95

5. Reframing the Modern Synthesis: Macroevolution | 130

6. Reframing the DNA Revolution | 165

7. Challenging the Modern Synthesis | 198

8. Rethinking the Religion–Science Relationship | 233

Bibliography | 265

Index | 277

Acknowledgments

TAKING A CONTROVERSIAL POSITION on a controversial issue leaves one with a limited number of conversation partners. I have been forced by circumstances to develop this work in considerable isolation. Nevertheless, the ideas here expressed have been inspired by interactions with many valued colleagues, even if those colleagues would not necessarily want to be associated with the conclusions I have reached. I will thank them anyway.

First, I would like to acknowledge Bruce Wrightsman, who though he is no longer with us, demonstrated to me early in my career the academic courage to take on controversial issues, and who shared with me an article he had written on Copernicus that challenged highly revered orthodoxies. Thanks also are due to Tex Sordahl, with whom I taught a course on ecology and religion for many years. Tex's lectures were really my first introduction to evolutionary theory and provided a foundation for my work in this book, even if I am now questioning aspects of what I learned from him. Of course, I would never have written this book had I not taken over the teaching of a science and religion course at Luther College taught by Loyal Rue. While I have taken the course in directions far different from Loyal's, his passion for this subject area has always been an inspiration to me. Finally, my biology colleague Eric Baack will find little to agree with here. But I appreciate his sharing with me articles of interest, even if his motive for doing so has been to try and talk me out of my conclusions!

I have also been immeasurably helped by the intelligent design (ID) community even if I am in some ways an outsider. Thanks are due to Michael Behe, who read early chapters and encouraged my completion of this work. Thanks also to John West for his interest in the project. And thanks to all those who have posted links to interesting articles and papers in various online venues that provided me with many important research leads. Many of these papers I would never have discovered on my own.

Thanks are due also to my home institution, Luther College, and especially the Dean's Office for supporting my 2014 sabbatical during which this book was conceived and during which much of the research was completed.

Special thanks to Bob Watson for always showing interest in my work and providing his own take on controversial issues. I always enjoy our conversations. And thanks to Tina, Amey, and Tyler for their interest in and support of my research and writing endeavors.

Finally, thanks are due to Wipf and Stock Publishers for taking on a project like this and making available to the reading public books from all different perspectives. They provide a valuable service in promoting ongoing conversation on many important issues. Any remaining errors are, of course, my own responsibility.

Introduction

I NEVER EXPECTED TO write this book. That I have become a critic of the Darwinian foundation of modern evolutionary theory comes as a surprise to no one more than me. Back when I began my academic career at Luther College almost twenty years ago, I had an interest in the relationship between religion and science owing to my undergraduate degree in science and engineering along with a PhD in religious studies. Teaching a class on the religion–science relationship would have been a natural fit, but it was not an option because such a course already existed and was taught by Loyal Rue, a senior faculty member with a number of publications in the field to his credit. I quickly learned that Luther College's Science and Religion course was Loyal's baby, so I developed my interests in other areas. But then Loyal retired, and knowing about my background in science and engineering, he encouraged me to take over responsibility for his baby and to continue to nurture an area of study that already had a long and illustrious history at Luther. I was more than happy to do so. But since the religion–science relationship was not my specialty in graduate school—nor had I followed the literature in the field for over a decade—I knew I had a lot of work to do to prepare to teach the course in a way that would be worthy of the tradition developed by Loyal (though the approach that eventually emerged is entirely different from his).

My scientific interests during my undergraduate days had revolved primarily around physics, astronomy, and cosmology; my interest in biology was minimal. But one cannot engage the religion–science relationship without squarely facing the acrimonious debate between evolution, creationism, and ID. Fortunately, at Luther I had team taught a course for a number of years with a colleague from the biology department on scientific and religious approaches to environmental issues. Through this experience I learned a great deal about modern evolutionary theory and scientific ecology, which provided me with a foundation on which to build as I began to read up on modern debates about evolutionary theory, creationism, and the ID movement. At that time, I fully accepted the idea that Charles Darwin had essentially solved the problem of the origin of species with his concept of natural selection, and that modern evolutionary theory was simply a more complex extension of Darwin's basic insight and stood as one of the most successful and empirically verified scientific theories ever proposed. But then I began to read the literature for myself and to my utter astonishment, I found I could no longer sustain this view. This transformation required a couple of steps.

As a trained biblical scholar, I could not—and still cannot—abide the arguments of young earth creationists, who read the creation story in Genesis literally and historically. (My earlier books have the titles *Was Jesus a Muslim?* and *Jesus and Jihad*, so my skepticism of Darwinism cannot be explained by appeal to conservative Christian ideology!) I was all but prepared to accept the popular wisdom calling intelligent design just a more sophisticated form of creationism—that ID proponents are just as motivated by religious concerns as their strict creationist brethren. In fact, I was all but prepared to accept without argument the common view that anyone who questions modern evolutionary theory is either unintelligent or a religious zealot—or both. But to teach the evolution/ID debate with integrity, I knew I needed to become more familiar with the ID literature. So with great hesitation I began to read it, beginning with Phillip Johnson's best-selling *Darwin on Trial* followed by books of Michael Behe and Stephen Meyer. Though I was not fully convinced by the arguments, I unexpectedly found this literature to be more interesting and scientifically substantive than the usual caricature. The ID literature raised serious questions about aspects of modern evolutionary theory that appeared to bear some weight. As any academic would, I became curious. If evolutionary theory is as empirically confirmed as most biologists say it is, and if ID can be easily disparaged and dismissed, why do biologists feel the need to create a caricature of the ID movement rather than simply facing it head-on and showing why it is wrong?

Something did not seem right, and I wanted to get to the bottom of it. I knew I needed to understand evolutionary theory much better than I did. So I spent a sabbatical immersing myself in the literature of the field. I traced the history of evolutionary theory from Darwin to Dobzhansky to Dawkins, reading a host of seminal volumes and papers, starting with the *Origin of Species* followed by the works of such luminaries as Ernst Mayr, Theodosius Dobzhansky, George Gaylord Simpson, James Watson and Francis Crick, Barbara McClintock, and many others. I pored over peer-reviewed papers appearing in such journals as *Science, Nature, Proceedings of the National Academy of Sciences, Evolution, Trends in Ecology and Evolution, Quarterly Review of Biology, American Scientist,* and many others. Through it all, I assumed that my more sympathetic feel for ID literature had to be due to my ignorance of evolutionary theory, and that once I had a better grasp of the latter, I would appreciate the problems with the former. Surprisingly, the opposite occurred.

Though I was reading scientific literature produced for a scientific audience, I read this literature from the perspective of a humanities scholar trained in the hermeneutics of suspicion. I was shocked when I began to recognize just how ambiguous and tentative so much of this literature is. It is littered with caveats, inconsistencies, unsupported assumptions, grand claims backed by a dearth of empirical evidence, and perhaps most surprising of all, by "religious" terms such as "orthodoxy," "heresy," "dogma," "creed," and "blasphemy." The more I read, the more confused I became, but at the same time the more confident I became that evolutionary biologists really do not know, a century and a half after Darwin, the mechanisms by which evolution occurs. It remains very much a mystery, as at least one biologist has admitted. I came to see that the claims made that Darwinian evolution is the most successful scientific theory of all time were just that—grand claims that constituted a grand narrative of Darwinian triumph. This grand narrative, I quickly learned, effectively obscures the very tentative nature of the scientific literature, thus advancing an ideological position in the guise of objective science. I had been taken in by this grand narrative as has anyone who has not taken the time to read the scientific literature in detail, which is most nonscientists. This would include many religion scholars, even those who deal specifically with the religion–science relationship. I realized that in a sense we had all been deceived by Darwin, and I set out to correct the record.

I have tried here to provide a service to religion scholars by taking the time to read the scientific literature at a depth that most other religion scholars understandably will not reach. It is normal to assume that more so-called popular books about evolution by such luminaries as Richard Dawkins, Daniel Dennett, and E. O Wilson provide accurate summaries of the current state of evolutionary theory so that their confident pronouncements about Darwinian evolution's explanatory power can be counted on to be correct. So it is not surprising that most religion scholars accept at face value this grand narrative of Darwinian triumph. My questioning of the veracity of this narrative will be deemed controversial—maybe even preposterous. But as we will see, even some highly trained evolutionary biologists have been deceived by Darwin too. Because scientists frequently rely on textbook presentations of scientific subjects rather than going back to read original papers, they too can fall prey to the way textbook discussions often oversimplify complex subjects, a point emphasized by Thomas Kuhn in *The Structure of Scientific Revolutions*. If evolutionary biologists can themselves be deceived by Darwin, it should be clear that I am not in any way intending to pass a negative judgment when I say that my colleagues in the religion field have been so deceived. This book is meant not as a critique of but a resource for those scholars—to provide them with a more honest look at the development of evolutionary theory unencumbered by the influence of the grand narrative of Darwinian triumph.

This book is also not meant to be a general attack on science or the scientific method. I have great respect for modern science and the wonders it has uncovered about the world in which we live. Questioning Darwinian evolution and proposing that this scientific theory acts more like a grand narrative of scientific triumph than an empirically confirmed scientific theory will sound controversial. But it is really nothing more than a recognition that science is just as much a human endeavor as religion, philosophy, politics, or economics. Scholars have been uncovering grand narratives in these other disciplines for a long time. The controversial claim would be that science has somehow escaped the influence of human bias and power dynamics. Of course it hasn't. And I will simply seek to make the influence of these power dynamics visible through a close reading and analysis of the literature of evolutionary biology, seeking to free religion scholars and others from the influence of this grand narrative in service to the development of new and creative approaches to the religion–science dialectic.

I apologize at the outset for the large number of direct quotes adorning this book. While I may have been able to improve the book's readability by paraphrasing more, I feel it is important, given the controversial nature of the argument, to hear the actual voices of the evolutionary biologists whose work leads to the conclusion that the mechanisms responsible for the origin and diversity of life truly remain a mystery.

These biologists do make that case, whether they intend to or not, and it is best to hear them in their own words as much as possible. How and why have we been deceived by Darwin? This is the troubling—but also fascinating—question I will seek to illuminate in the pages that follow.

1

Darwinian Biology's Grand Narrative of Triumph and the Subversion of Religion

LEHIGH UNIVERSITY BIOCHEMIST MICHAEL Behe thrust himself into the center of public controversy in 1996 with the publication of *Darwin's Black Box: The Biochemical Challenge to Evolution*, a best-selling book that quickly became a foundational text of the widely disparaged intelligent design (ID) movement. In a surprisingly sympathetic review, University of Chicago biologist James Shapiro commented: "*Darwin's Black Box* has the merit of showing us that evolution remains a mystery. Its fundamental driving forces have not been resolved either in detail or in principle."[1] That a biologist connected to such a prestigious university would openly sympathize with the views of an ID advocate and seek to undermine certainty in our understanding of the evolutionary process is truly startling; Shapiro's characterization runs counter to virtually the entirety of the vast corpus of literature produced in the academic field of evolutionary biology.

Rather than viewing evolution as a continuing mystery more than a century and a half after the publication of Charles Darwin's *On the Origin of Species*, we are more likely to hear that Darwin's theory "may prove to be the greatest revolution in the history of thought"[2] and "as close to truth as any science is ever likely to get."[3] According to the great Harvard evolutionary biologist Ernst Mayr, "The theory of evolution is quite rightly

---

1. Shapiro, Review of *Darwin's Black Box*.
2. Ghiselin, *Triumph of the Darwinian Method*, 1.
3. Barash, *Natural Selections*, 22.

called the greatest unifying theory in biology."[4] Darwinian philosopher Daniel Dennett is even more effusive, calling Darwin's theory "the single best idea anyone has ever had,"[5] an idea that promises "to unite and explain just about everything in one magnificent vision."[6] In Dennett's view, Darwin's theory is now beyond dispute among scientists. "It unifies all of biology and the history of our planet into a single grand story."[7]

In light of these grand claims about the explanatory power of Darwinian evolution, how should we understand Shapiro's more circumspect characterization of evolution as a continuing mystery? Can we simply dismiss Shapiro as a marginal voice? Contrary to the confident pronouncements about evolutionary theory's virtual factuality issuing from the likes of Mayr, Dennett, Richard Dawkins, and many others, I will defend the unlikely thesis that Shapiro's characterization represents a far more accurate picture of the current state of evolutionary theory as it emerges from a critical engagement with the peer-reviewed scientific literature. I know such a statement, made as it is by a religion scholar, will sound hopelessly naïve, and I will be accused of inappropriately stepping far outside my disciplinary boundaries. But the real world does not come to us in the hermetically sealed packages of modern academic disciplines, and there is no reason why reading scientific literature from the perspective of a religion scholar cannot yield valid insights into the question of life's origin and diversity. Besides, we do not need to venture far into the literature of evolutionary biology to see that grand assertions about evolutionary theory's explanatory power do in fact begin to crumble under the weight of critical scrutiny.

Consider Ernst Mayr, for example. He accuses dissenters from Darwin's theory of displaying such a colossal ignorance of the evolutionary biology literature that refuting them would be a waste of time. The essential features of the modern theory of evolution are so "consistent with the facts of genetics, systematics, and paleontology that one can hardly question their correctness."[8] Mayr characterizes the basic theory of evolution as "a two-stage phenomenon: the production of variation and the sorting

---

4. Mayr, *Animal Species and Evolution*, 1.
5. Dennett, *Darwin's Dangerous Idea*, 21.
6. Ibid., 82.
7. Ibid., 20.
8. Mayr, *Animal Species and Evolution*, 8.

of the variants by natural selection."[9] But agreement on this basic thesis does not mean that work in evolutionary biology is complete. The basic theory, according to Mayr, is in many cases "hardly more than a postulate and its application raises numerous questions in almost every concrete case."[10] Modern research is directed at evolutionary phenomena that do not yet appear to be explained by the current theory. This leads to points of conflict between biologists over the interpretation of anomalous data. But Mayr stresses that "none of these arguments touches upon the basic principles of the synthetic theory. It is the application of the theory that is involved, not the theory itself. And with respect to application we still have a long way to go."[11]

Consider Mayr's words carefully. In the same place where he confidently asserts the essential factuality of Darwinian evolution, he also characterizes this theory as "hardly more than a postulate," and admits that the application of this theory to actual biological phenomena "raises numerous questions in almost every concrete case." Mayr's confident assertion seems betrayed by his own frank admission that fundamental aspects of the evolutionary process at the time he was writing remained unclear and without empirical verification. No scientific theory that is "hardly more than a postulate" and whose application "raises numerous questions in almost every concrete case" deserves at the same time to be considered "as close to truth as any science is ever likely to get."

But Mayr was writing in 1965. Perhaps the situation has improved with the great advances made in biology over the last fifty years. Evolutionary theory surely now stands on firmer ground. Perhaps, but consider the website sponsored by the biology department at the University of California, Berkeley titled Understanding Evolution. This website provides the general public with an up-to-date summary of the current state of evolutionary theory. On it we read, "All available evidence supports the central conclusions of evolutionary theory, that life on Earth has evolved and that species share common ancestors. Biologists are not arguing about these conclusions. But they are trying to figure out how evolution happens, and that's not an easy job."[12] Among the questions that evolutionary biologists are trying to answer, according to the Berkeley website,

9. Ibid., 8.
10. Ibid.
11. Ibid., 9.
12. "Understanding Evolution."

is, "How does evolution produce new and complex features?" This would seem to be *the* central question of evolution. We cannot confidently declare the essential factuality of evolutionary theory if we do not know how to account for the development of the new and complex features that lead to the evolution of whole new kinds of organisms. If biologists are still trying to answer this question, the state of evolutionary theory seems not to have advanced significantly since 1965 when Mayr was writing, and assertions of evolutionary theory's virtual factuality appear to be overstatements. Shapiro's more circumspect evaluation of the state of our knowledge may be closer to the truth. But then what accounts for the triumphant assertions of evolutionary theory's explanatory power so often encountered in literature designed for the wider public?

Since the early twentieth century Darwinian evolution has been perpetuated in the form of a grand narrative of triumph that consistently overstates the empirical validity of its Darwinian framework: that evolution primarily occurs as a result of natural selection acting on the inherent variability of organisms. I will read the literature of evolutionary biology from the perspective of a religion scholar trained in the hermeneutics of suspicion in order to demonstrate that the history of evolutionary theory's development is better read as a continuing saga of contestation over fundamental, still unresolved, questions, not as a history of how one theory came to triumph over all others. The grand narrative of Darwinian triumph, like all grand narratives, turns out to be an ideologically motivated story that, ironically, is consistently undermined by the very scientific work designed to promote it.

But why is it important for scholars of religion, and particularly those interested in the intersection between science and religion, to recognize the ideological nature of evolutionary biology's grand narrative of Darwinian triumph? Because the biological establishment has been so effective at marketing this narrative that almost everyone from outside the world of the biological sciences has come to uncritically accept it, and this includes scholars of religion. In fact, intellectual adherence to Darwinian evolution has itself become something of a litmus test for intelligence. Anyone who dares question it is considered to be either woefully ignorant or a religious fanatic—or both. Or as Daniel Dennett put it, "those whose visions dictate that they cannot peacefully coexist with the rest of us we will have to quarantine as best we can."[13]

---

13. Dennett, *Darwin's Dangerous Idea*, 519.

Not wishing to be quarantined, religion scholars and theologians who wade into the troubled waters of the religion–science debate have clearly felt compelled to uncritically accept Darwinian evolution and base their reflections about the relationship between religion and science on this Darwinian foundation. In exercising this deference to science, I will argue that such scholars have unwittingly been "deceived by Darwin." If the strength of Darwinian evolution turns out to be more tenuous then the grand narrative portrays, it may not be necessary for religion scholars and theologians to be the ones always capitulating to the scientists. Religion scholars may have something more constructive to contribute to this debate.

In the balance of this chapter, I will provide examples demonstrating how religion scholars, theologians, and in some cases even scientists themselves have been deceived by Darwin by uncritically accepting the grand narrative of Darwinian triumph. Then I will provide a summary of this grand narrative, followed by a consideration of its ideological purpose, and finally I'll conclude with a summary of a more complex and nuanced counternarrative to be developed in more detail in the balance of this book. It may not be necessary for religion scholars, theologians, or anyone else, for that matter, to continue to be deceived by Darwin.

Deceived by Darwin

The general acceptance of Darwinian evolution has been standard practice for scholars of religion for a long time. Even critic Taner Edis can write, "Many theologians have become too deferential to modern science."[14] This is not surprising given that most religion scholars understandably possess neither the time nor the inclination to read deeply into the peer-reviewed scientific literature. I will consider a few examples here to demonstrate how such acceptance may be leading these scholars astray.

I begin with Ian Barbour, one of the most influential scholars of the religion–science dialectic in the latter part of the twentieth century. In his 1997 book *Religion and Science: Historical and Contemporary Issues*, Barbour begins a chapter on evolution noting that scientists have accumulated immense quantities of evidence to support, both that evolution has occurred, and that the Darwinian mechanism of natural selection

---

14. Edis, "Religion: Accident or Design?" 389.

acting on the inherent variability of organisms is the main driver of evolutionary change. He concedes the vigorous debate among scientists over the details of the theory and the possible role of factors beyond natural selection, but it is clear that he accepts the general form of Darwinian evolution as it is advanced by the biological establishment.[15]

This is confirmed when Barbour cites the phenomenon of industrial melanism in moths as evidence of natural selection in action. *Industrial melanism* refers to the darkening of lighter-colored moths in England during the nineteenth century as coal soot from the industrial revolution darkened tree trunks making lighter-colored moths more conspicuous and therefore easy prey for birds. The darker-colored moths resulted from a genetic mutation, and during the industrial revolution these mutant forms were more likely to survive and pass their genes on to the next generation. As Barbour says, "on the soot-darkened trees of industrial areas, the dark moth is less conspicuous; in the past century it has completely supplanted the light-colored form in parts of England."[16] As Barbour notes, after antipollution legislation was passed in the twentieth century, tree lichens became lighter in color again, and over time the light-colored moths came to once again dominate the population. This appears to be a clear case of natural selection at work on time scales easily observed by humans, and industrial melanism in the peppered moth has become literally a textbook case of observable natural selection in action.[17]

Barbour seems unaware, however, that the scientific literature on industrial melanism is far more ambiguous than the portrayal he has absorbed from the grand narrative of Darwinian triumph. He displays no awareness of the controversy that surrounded a series of famous experiments performed by Bernard Kettlewell in the 1950s designed to empirically verify the bird predation theory of industrial melanism and put this example of natural selection on firm empirical grounds. As we will see in more detail in chapter 4, biologists continue to debate whether the phenomenon of industrial melanism is in fact a clear example of natural selection, and alternative, non-Darwinian explanations exist to explain the phenomenon. Moreover, these debates over the phenomenon of industrial melanism show clear signs of being driven by larger, ideological

---

15. Barbour, *Religion and Science*, 221.

16. Ibid., 222.

17. The peppered moth example can be found in virtually every textbook published about evolution as well as in other literature designed to convince the general public of the truth of Darwinian evolution.

concerns. To the extent that Barbour uncritically accepted industrial melanism as a clear example of natural selection in action, he was clearly falling prey to the distortions of the grand narrative.

In a similar manner, Barbour takes issue with those who argue that extremely intricate organs like the human eye could not have evolved through the Darwinian mechanism of random variation and natural selection. He cites Hugh Montefiore's defense of God as the most probable explanation for the intricacies of design found in the eye and other complex organs. Barbour, however, considers Montefiore's argument a sophisticated form of the "God of the gaps" explanation and writes, "It is vulnerable insofar as these gaps in the scientific account have or will be filled in."[18] Once again, Barbour has accepted an idea aggressively pushed within the grand narrative of evolutionary theory: that the evolution of intricate organs like the eye can be explained by Darwinian mechanisms. But as we will see in chapter 2, the scientific literature tells a considerably more ambiguous story about the evolution of the eye and the ability of a Darwinian mechanism to account for the extreme intricacy of this organ of vision that even Darwin admitted gave him a cold shudder.

Given Barbour's uncritical acceptance of standard evolutionary theory, especially the idea that biologists have succeeded in showing how a purely naturalistic process can account for the great diversity of life, he clearly will not be able to accept traditional religious understandings of God as creator. He therefore advocates a "theology of nature" that, he says, "is based primarily on religious experience and the life of the religious community but which includes some reformulation of traditional doctrines in the light of science."[19] This idea that religious ideas must always be reformulated in the light of science—never the reverse—is pervasive in literature addressing the religion–science dialectic, and it is on full display in the work of an influential Catholic theologian who takes science seriously.

John Haught has written extensively on the issue of reformulating Christian theology in the light of science, particularly the science of evolutionary biology. In two influential books, *God after Darwin: A Theology of Evolution* (2000) and *Deeper Than Darwin: The Prospect for Religion in the Age of Evolution* (2003), Haught repeatedly affirms his acceptance of Darwinian evolution:

---

18. Barbour, *Religion and Science*, 246.
19. Ibid., 247.

8   THE MYSTERY OF EVOLUTIONARY MECHANISMS

> My own preference is to concede from the outset the general integrity of Darwinian *science* (which I would distinguish carefully from the materialist ideology in which it is often packaged). Certainly Darwinian ideas are not perfect. That evolutionary theory will continue to undergo revision, I have no doubt. Even today there is some discontent with neo-Darwinism in the scientific community. But out of respect for the thinking of most contemporary scientific experts, and especially the majority of biologists, in this book I shall take Darwin's version to be a reasonably close, though incomplete and abstract, approximation of the way life has developed on earth.[20]

Haught calls *On the Origin of Species* "one of the most important works of science ever written," accepting that "recent scientific expertise seems to confirm its explanatory power."[21]

The essential correctness of Darwinian evolution becomes then the foundation for Haught's theological reflections. If Darwin was right as Haught believes, then, like Barbour, Haught argues that traditional ideas about God as creator must be abandoned. We must be willing to give up comfortable ideas about divine order and reconceive God, not as a creator who brings the universe into existence through direct divine action, but rather as a spirit that draws the universe forward into a future that itself is not divinely determined. Haught accepts the concept of contingency in evolution made popular by Stephen Jay Gould—that if the evolutionary tape were rewound and played forward again, the story would come out differently. It is not even clear to Haught that humans would be the result of a new evolutionary scenario. Nevertheless, he is convinced that a Darwinian view of evolution "answers to the deepest intuitions of religion."[22] It is not just that Haught sees religion and Darwinian evolution as compatible; he understands Darwinian evolution to be the very foundation for a religious view of life. As a practicing theologian who accepts the essential truth of Darwinian evolution and who wants to be taken seriously by the academic community, Haught has little choice but to figure out a way to base his theological outlook on the foundation of Darwinian evolution. This is fine if Darwinian theory is as well established scientifically as Haught thinks it is. But if it isn't, Haught's uncritical acceptance of Darwinian evolution may actually undermine his theological work;

---

20. Haught, *God after Darwin*, 14 (italics original).
21. Haught, *Deeper Than Darwin*, 69.
22. Haught, *God after Darwin*, 6.

it may not be necessary to entirely forsake more traditional theological ideas about God as creator. The grand narrative of Darwinian triumph may therefore be subverting religion in Haught's otherwise thoughtful work.

More recently *Zygon*, an academic journal dedicated to the religion–science relationship, has published a series of articles arguing against ID as an adequate way to conceive the relationship between religion and science. Gloria Schaab, in an article summarizing the work of Arthur Peacocke, terms the relationship between religion and science one of mutually illuminative interaction in which insights from each discipline inform the understandings and discourse of the other. Yet she concludes that theology must be reconceived in light of an evolutionary worldview.[23] It is unclear how or whether an evolutionary worldview must be reconceived in light of theology, raising the question of how mutually illuminating such dialogue can be. Similarly, in an article provocatively titled "Could God Create Darwinian Accidents?" John Wilkins states, "it is important to me that intelligent people of a religious bent are able to accept the results of modern science without needing to modify it to suit their religious doctrines, which leads ultimately to a subordination of science to religion."[24] The implication here is that religion should always be subordinated to science, and that religious people who would resist this subordination might be lacking in intelligence.

Articles like this appearing in the pages of *Zygon* repeatedly accept the results of modern evolutionary theory as the foundation for religious reflection, a stance that allows the journal to be viewed as intellectually responsible and academically acceptable. According to the "statement of perspective" found on *Zygon*'s website:

> Traditional religions, which have transmitted wisdom about what is of essential value and ultimate meaning as a guide for human living, were expressed in terms of the best understandings of their times about human nature, society, and the world. Religious expression in our time, however, has not drawn similarly on modern science, which has superseded the ancient forms of understanding. As a result religions have lost credibility in the modern mind.[25]

---

23. Schaab, "Evolutionary Theory and Theology."
24. Wilkins, "Could God Create Darwinian Accidents?" 39.
25. *Zygon: Journal of Religion and Science*, "Who We Are."

For the editors of *Zygon*, religious thought has credibility only when it is based on an uncritical acceptance of the results of modern science—including modern evolutionary theory. But this subordination of religion to science unfortunately comes at the cost of not promoting an authentic dialogue between religion and science. Once again, religion is subverted by the grand narrative of Darwinian triumph.

Most recently this theme emerges in a dialogue about religion and evolution between Michael Ruse, a noted Darwinian atheist philosopher, and Michael Peterson, an evangelical Christian philosopher. While we would expect Ruse to be firmly committed to Darwinian evolution, we may be surprised at the level to which Peterson shares this commitment. For example, we are told by Peterson that Darwinism is one of the greatest achievements of intellectual history, and indeed that Darwin *discovered* natural selection.[26] This last point deserves further comment. Even the grand narrative of Darwinian triumph admits that Darwin merely developed natural selection as an explanatory hypothesis, one that required more than a century of continuing empirical work to confirm. In no sense did Darwin *discover* natural selection as one discovers a new species previously unknown to science. By characterizing natural selection as a Darwinian *discovery* rather than a Darwinian *hypothesis*, Peterson demonstrates how steeped he is in the grand narrative of Darwinian triumph, with its tendency toward Darwinian hagiography.

And he does not disappoint: "As a theory in science, evolution is as well confirmed as any empirical theory could possibly be, providing the most reasonable interpretation of a myriad of facts in a wide variety of disciplines, such as paleontology, comparative anatomy, biogeography, biochemistry, embryology, and ecology."[27] This really is a gross mischaracterization of evolutionary theory. As we will see, the fossil record has always stood as a major obstacle to Darwinian theory, as even Darwin himself recognized. Some of the biggest critics of Darwin throughout history have been paleontologists. But Peterson is undeterred. Reminding his readers that he and Michael Ruse both stipulate the validity of evolutionary biology, he asks, "How could we not? Theodosius Dobzhansky, a famous Russian biologist and Orthodox Christian, published an article entitled 'Nothing in biology makes sense except in light of evolution.' The defensive ploy stating that 'evolution is a *theory*, not a *fact*' badly

---

26. Peterson and Ruse, *Science, Evolution, and Religion*, 105.
27. Ibid., 109.

misunderstands that well-confirmed theories in science are considered facts."[28] Richard Dawkins could not have said it any better!

How, then, does Peterson square his dogmatic defense of Darwinian evolution with his evangelical Christian worldview? He cites John Haught's idea that Darwin bequeathed a great gift to theology. If the truth of natural selection is so highly confirmed, then according to Peterson, "it is much wiser—and much more faithful to a classical theological perspective—to provide a positive philosophical interpretation of it . . . After all, if theology is to be relevant and meaningful, it must freshly articulate enduring truths in light of emerging knowledge in every generation."[29] For Peterson some aspects of God's ways are understood more clearly in light of evolutionary truth. For example, strong views of divine control over creation must give way to a more dynamic conception in which "God's role is generally supportive and interactive but not always determinative."[30] Religion, Peterson argues, has no business making pronouncements about "the empirical mechanisms at work in God's world. Theology should concern itself with expressing broad principles regarding God's love and plan for humans while affirming ongoing scientific inquiry into the created world."[31] Peterson has bought into Stephen Jay Gould's idea of religion and science as constituting nonoverlapping magisteria. But by disconnecting religious reflection from the world of material existence, has Peterson rendered the God he worships unreal, a mere figment of his imagination? I will take up this question in detail in this book's final chapter. Peterson clearly feels the need to build his Christian theological worldview on the foundation of Darwinian evolution in order for his theological work to be seen as relevant and intellectually respectable. But in so doing, religious reflection is actually being subverted by a scientific grand narrative.

To appreciate just how pervasive (and in some ways how perverse) Darwinian ideology has become, we need only consider the recent development of evolutionary approaches to the explanation of religion. Since the foundation of sociobiology by E. O. Wilson in the 1970s, a veritable army of biologists, psychologists, psychiatrists, and anthropologists have been busy trying to explain the very existence of religion as

---

28. Ibid.
29. Ibid., 110.
30. Ibid., 112.
31. Ibid., 111.

an evolutionary phenomenon. Religion, we are told, made early human societies better adapted to their conditions of life (by fostering cooperation over competition, for instance), and religion was therefore naturally selected and passed down to succeeding generations. This application of evolutionary theory to religion is exemplified in Dean Hamer's best-selling book *The God Gene: How Faith Is Hardwired into Our Brains*, in which we are told that mystical experiences like the apostle Paul's on the road to Damascus might well be scientifically explained as temporal lobe epileptic seizures.[32] Unfortunately, Hamer seems blissfully unaware of the fruits of modern biblical scholarship, which understands the Damascus Road story to be a literary creation of the late first-century writer of the book of Acts, not the historical recounting of Paul, who describes his own conversion experience in very ordinary terms in his letter to the Galatians. Likewise, in evolutionary psychologist Ara Norenzayan's *Big Gods: How Religion Transformed Cooperation and Conflict*, we are told that belief in supernatural beings constitutes "the hallmark of all religions."[33] Norenzayan has universalized a particularly Protestant Christian view of religion and left Buddhism, Confucianism, and secular Judaism from consideration as religions!

Recently, I attended the Second International Conference on the Evolution of Religion held at a beautiful resort on Native American land in New Mexico. The high desert surroundings were spectacular, but the content of the sessions was in some cases nothing short of absurd. One presenter claimed to have worked out a complete phylogenetic tree of the world's religions, showing how they evolved and branched off in an orderly way from a hypothetical original primitive religion (but she cited no literature from the history of religions to ground her analysis)! Another presenter built on Dean Hamer's thesis and tried to explain the development of all religions by an appeal to the experience of temporal lobe epilepsy of their reputed founders—Jesus, Paul, Abraham, Moses, Muhammad, Buddha, and so forth. The evolution-of-religion discourse has unfortunately remained hermetically sealed off from any influence from humanities-based religion scholarship, and this is likely intentional. For religion to yield to an evolutionary explanation, it must be understood as having a universality and homogeneity about it since humans share a common biological endowment. By problematizing

---

32. Hamer, *God Gene*, 132.
33. Norenzayan, *Big Gods*, 7.

universalizing discourses and emphasizing the complexity and cultural particularity of religious beliefs and practices, humanities-based religion scholarship renders a one-size-fits-all evolutionary explanation virtually impossible. Such scholarship is therefore ignored, replaced by the modern materialist demand to explain everything in Darwinian terms. In the words of the Evolution Institute, a think tank that advocates this viewpoint, "the Evolutionary Institute is capable of applying a single theoretical framework—evolution—to all major policy issues at all scales, from neighborhoods to nations to world history."[34] These scientists appear to be deceiving themselves by their own grand narrative. For if it turns out that Darwinian evolution cannot even account for the origin of species, it certainly will not be able to account for the origin of religions.

We no longer need to be deceived by this Darwinian grand narrative. We must instead wrestle fully with the ambiguous nature of modern evolutionary theory as it is characterized by James Shapiro, and open ourselves to the possibility that religious ideas may have a constructive role to play in how we think about the origin and diversity of life. What, then, is the central plot of this grand narrative of Darwinian triumph that has deceived so many? And how does it function?

The Grand Narrative of Darwinian Triumph

No one has done more to perpetuate the grand narrative of Darwinian triumph than the great Harvard biologist Ernst Mayr. I will, therefore, use the preface to the 1998 edition of the seminal volume he coedited with William Provine, *The Evolutionary Synthesis: Perspectives on the Unification of Biology*, as a framework for this summary. As one of the chief architects of the so-called evolutionary synthesis of the 1940s, Mayr, who lived to the age of 100 (and was still writing books in his nineties), continued to be one of the most outspoken of Darwinian apologists right up to the end of his long life.

The story of the development of modern evolutionary biology begins with Charles Darwin and his publication of *On the Origin of Species* in 1859, in which Darwin proposed the concept of natural selection as a naturalistic mechanism to explain the evolutionary process. In contrast to Daniel Dennett's ascribing to Darwin one of the greatest ideas anyone has ever had, or Michael Peterson's view that Darwin *discovered* natural

---

34. Evolution Institute, "What We Do."

selection, most evolutionary biologists will admit that Darwin largely failed to convince his contemporaries of the truth of his theory. With a few exceptions, most of Darwin's nineteenth-century interlocutors could not conceive of how gradual changes brought about by natural selection acting on the inherent variability of organisms could lead to the creation of whole new varieties of life. And they certainly did not think Darwin had provided compelling evidence to document this. The period from the late nineteenth century until about 1920 is often called the eclipse of Darwinism.

One of the major problems with Darwin's theory was its ignorance of the laws of heredity. Darwin knew from his own painstaking observations that individual organisms of the same species differ from one another in innumerable ways. But he did not know how variations were produced or how they were passed on from one generation to the next. This began to change, however, in 1900 with the rediscovery of the work of the Moravian monk Gregor Mendel. Mendel had worked out rules of inheritance in his famous experiments with pea plants by the 1860s, but his work lay largely unknown until the Dutch geneticist Hugo de Vries brought it into the growing scientific mainstream. Rather than provide support for Darwinism, however, de Vries along with other early geneticists such as William Bateson understood Mendel's work as posing a challenge to Darwin's theory. Darwin was committed to a view of evolution as occurring gradually through the accumulation of small changes over long spans of time—the doctrine known as *natura non facit saltum* (nature does not make jumps). But de Vries, Bateson, and others, who became known as the Mendelians, took the opposite view: that evolution could only occur as the result of major evolutionary jumps, or saltations. This non-Darwinian saltational view came to dominate evolutionary thinking in the early twentieth century.

For his part, Mayr critiques the Mendelians as being proponents of what he calls typological thinking. That is, the Mendelians viewed species as ideal types with strict lines of separation between them. Evolutionary change could therefore only occur in large jumps as one ideal type transformed into another. The lack of truly transitional forms in the fossil record in the early twentieth century only seemed to reinforce this view. But Mayr praises Darwin for replacing this typological thinking with what he calls population thinking. For Darwin, species don't evolve into new species. Rather, populations of organisms evolve gradually over time as natural selection works on the inherent variability of individual organisms. Since the individual organisms that make up a species vary so much from one another, Darwin reasoned that species could not be understood as representing ideal types. Evolution, for Darwin, was a phenomenon that operated at the level of populations, not individual organisms.

By the 1920s, a number of biologists began to take population thinking seriously, developing models to understand how small variations among the individuals in a population could be passed on from one generation to the next in such a way that natural selection could bring certain traits into dominance within a population. This is the field known as population genetics, founded by three mathematically gifted individuals—Ronald Fisher and J. B. S. Haldane in Great Britain, and Sewall Wright in the United States. Together, they worked out sophisticated quantitative models demonstrating how small genetic changes could spread through a population in such a way as to drive gradual evolutionary change. The center of gravity for evolutionary thinking thus began to shift away from the saltational thinking of the Mendelians and toward the Darwinian population thinking of the population geneticists. But a problem still remained. The population geneticists studied the genetic changes occurring only in local gene pools and were not in a position to consider larger issues of organic diversity and macroevolutionary change. These topics were the province of naturalists and taxonomists. Mayr sums up:

> The result was the existence of two camps of evolutionists: the population geneticists, who provided a correct picture of the genetic changes in a given local population but were ignorant about the origin of organic diversity, and the naturalists (taxonomists), who understood species and speciation but were still fighting the essentialistic Mendelians, being unaware of the recent advances in genetics.[35]

Mayr has set us up for the introduction of a hero who will bring all the pieces together.

Enter Theodosius Dobzhansky, a naturalist who emigrated from Russia to the United States in 1927 to learn genetics in the laboratory run by Thomas Hunt Morgan at Columbia University. In Mayr's view, the gap between the population geneticists and the naturalists was bridged by Dobzhansky in 1937 with the publication of his seminal text *Genetics*

---

35. Mayr, "Preface, 1998," xii.

and the Origin of Species. In this book, Dobzhansky demonstrated that "there was no conflict between the findings of the two camps, and that each camp needed the findings of the other for a full understanding of evolution."[36] This message was enthusiastically received by biologists in both camps, and over the next decade, this message was elaborated further and cemented into place by the likes of Mayr himself (working on species and speciation), George Gaylord Simpson (paleontology), Bernhard Rensch (zoology), and George Ledyard Stebbins (botany). In 1947, a symposium was held at Princeton University where representatives from the diverse branches of biology "came together and declared their acceptance of this newly synthesized, integrated evolutionary theory."[37]

The period from the late 1930s until about 1950 was dubbed "the modern synthesis" by Julian Huxley (grandson of T. H. Huxley) in his foundational text *Evolution: The Modern Synthesis*. Despite the fundamental disagreements dividing evolutionary thinkers in the early 1900s, Mayr states, "within the short span of twelve years (1936–1947), the disagreements were almost suddenly cleared away and a seemingly new theory of evolution was synthesized from the valid components of the previously feuding theories."[38] Of course, Mayr cannot deny that criticisms of Darwinian theory continued to emerge in the decades following the grand synthesis, but he argues, "None of the numerous efforts of the past sixty years to refute the synthesis has been successful."[39] The synthesis, for Mayr, was nothing but a confirmation of Darwin's original theory. We can rest assured that "the basic principle of Darwinism is as solid at the end of the twentieth century as it was in 1859, untouched by all the advances made by evolutionary biology and other branches of biology in the intervening years."[40]

The marks of a grand narrative that greatly oversimplifies a long and complex history are unmistakable here: the building of a narrative tension between the population geneticists and the naturalists; the resolving of that tension by the introduction of the hero Dobzhansky, who creates the grand synthesis; the falling in line around Darwinism by the various branches of biology over the brief period of a decade; and the confident

---

36. Ibid.
37. Ibid.
38. Mayr, "Preface to the Original Edition," xv.
39. Mayr, "Preface, 1998," xiii.
40. Ibid.

assertion that the basic principle of Darwinism has not been touched by the extraordinary advances made in biology since the 1950s. In 1984, Ernst Mayr wrote a reflection for the *Times Literary Supplement* titled "The Triumph of Evolutionary Synthesis"[41] to recount and celebrate the triumphant nature of the modern synthesis of evolutionary theory. And in a 1976 essay, he referred to "an all-powerful natural selection."[42] Recall, however, that Mayr in other places calls Darwinian evolution "hardly more than a postulate," admitting that the application of this theory to actual biological phenomena "raises numerous questions in almost every concrete case." So we must consider the narrative of Darwinian triumph with appropriate skepticism. What is the true purpose of this narrative if it provides an overly simplistic account of the development of evolutionary theory?

Ideology and the Grand Narrative

Grand narratives develop for a reason. They intentionally oversimplify the vagaries of complex histories in the interest of advancing an ideological agenda. So we must ask, What ideological agenda stands at the center of the grand narrative of Darwinian triumph? In many ways, the rest of this book will be an answer to this question, but it will be helpful here to provide a tentative answer in brief outline.

In short, evolutionary biology's grand narrative of Darwinian triumph serves to shore up the status of biology as a proper naturalistic science. Prior to the twentieth century, biology (or natural history as it was called at the time) was bound up with religion in a much more significant way than virtually any other of the emerging scientific disciplines. Many natural historians of the eighteenth and nineteenth centuries were clergy, especially in England, and the idea that the various forms of life had each been uniquely created by God in recent history was the reigning orthodoxy for most people. While Darwin was not the first person to question this idea, he is the one who based his evolutionary theory on the foundation of detailed empirical observation while also proposing a mechanism of evolutionary change—natural selection—that could account for evolution without recourse to any kind of supernatural, intelligent, or vitalistic force. As Ernst Mayr frames it:

41. Mayr, "Triumph of Evolutionary Synthesis."
42. Mayr, *Evolution and the Diversity of Life*, 3.

> The Darwinian revolution was not merely the replacement of one scientific theory by another, as had been the scientific revolutions in the physical sciences, but rather the replacement of a worldview, in which the supernatural was accepted as a normal and relevant explanatory principle, by a new worldview in which there was no room for supernatural forces.[43]

Darwin set the stage for biology to emerge from its roots in natural theology, to break loose from its theological underpinnings, and to aspire to become a rigorous, empirical, and fully naturalistic science like physics and chemistry.

This may seem a heavy burden to place on the shoulders of one dyspeptic nineteenth-century amateur scientist with a controversial theory, but non-Darwinian biologist Michael Denton has put it best:

> Undoubtedly, one of the major factors which contribute to the immense appeal of the Darwinian framework is that, with all its deficiencies, the Darwinian model is still the only model of evolution ever proposed which invokes well-understood physical and natural processes as the causal agents of evolutionary change. Creationist theories invoke frankly supernatural causes, the Lamarckian model is incompatible with the modern understanding of heredity . . . and saltational models of evolution can never be subject to any sort of empirical confirmation. Darwinism remains, therefore, the only truly scientific theory of evolution. It was the lack of any scientific alternative which was one of its greatest attractions in the nineteenth century and has remained one of its enduring strengths ever since 1859. Reject Darwinism and there is, in effect, no scientific theory of evolution.[44]

That Denton is correct is beautifully illustrated by two passages from the recently published *Princeton Guide to Evolution*, a compendium of articles on all the latest research in evolutionary biology. In an article on natural selection, evolutionary biologist Douglas Futuyma calls Darwin's idea one of the most important ideas in the history of the world because:

> for the first time, there existed a purely scientific explanation for the most powerfully impressive examples of design and purpose in nature: the features of organisms that equip them most exquisitely for survival and reproduction. The adaptations

---

43. Mayr, "Evolution and God," 285.
44. Denton, *Evolution: A Theory in Crisis*, 355.

of organisms—including humans—had long been attributed to the Creator's beneficent design, indeed were among the most important arguments for the existence of such a supernatural Creator. As an alternative explanation lacking the slightest supernatural tinge, natural selection made biology a science.[45]

Futuyma, interestingly, does not call natural selection one of the most important ideas in the history of the world because of its empirically verified truthfulness, but primarily for its role in making biology a science! He is not alone. In the same volume, Stephen Stearns writes:

> Darwin's greatest idea, natural selection, explains the evolution of the myriad of such astonishing adaptations through the operation of simple mechanisms that can be readily observed. A triumph of human thought, it organizes and explains much of biology, strongly contributes to the impact of biology on other disciplines, and as a major scientific principle not contained in chemistry or physics, elevates biology to their rank in its power to explain the natural world.[46]

For Futuyma and Stearns, the importance of natural selection is found in its power to establish biology as a science and to raise biology to the status of chemistry and physics. It is truly telling that neither argues for the empirically verified truthfulness of the theory of natural selection as its most important attribute!

Advancing Darwin's theory as the triumphant theory of evolution becomes the primary way the modern scientific discipline of the biological sciences maintains its fully scientific status. The grand narrative of Darwinian triumph then primarily serves the guild interests of the biological establishment. Whether Darwinian theory is true or not becomes less important than its role in reinforcing these guild interests. We see evidence of this in the way evolutionary biologists review the work of ID theorists, the most vocal and organized contemporary critics of Darwinian theory. The biological establishment repeatedly argues that ID is not a scientific theory but merely a more sophisticated form of religious creationism. Yet, ironically, members of the biological establishment cannot seem to resist the urge to raise the scientific status of ID by reviewing the works of these "nonscientists" in the peer-reviewed scientific literature.

---

45. Futuyma, "Natural Selection and Adaptation," 189.
46. Stearns, "Natural Selection, Adaptation, and Fitness," 193.

They clearly view ID as a significant threat. But what kind of threat? Consider a few examples.

In an article titled "Intelligent Design Creationism: None of Your Business? Think Again," Barry Palevitz argues that if the ID people have their way, "the prospects for science as currently practiced are far direr than many reading this essay can conjure."[47] The paleontologist Lisa Park is even more explicit in an article titled "It's Not about Evolution: The Debate about Intelligent Design." For Park this debate is about the very future of scientific infrastructure in the United States. The U.S. is slipping as a world leader in science and attacks on evolution are a measure of the hostility society has toward science in general. Maintaining the integrity of science is important to all of society, not only to evolutionary biologists. The future of scientific funding may be at stake.[48] Perhaps the most telling response comes from Marshall Berman. In an article titled "Intelligent Design Creationism: A Threat to Society—Not Just Biology," Berman sets the tone for his essay with the famous quote ascribed to Edmund Burke, "The only thing necessary for the triumph of evil is for good men to do nothing." Berman then launches into a polemical attack on ID, accusing its proponents of pushing such a nefarious religious agenda that "It is time for those who cherish our republic and our freedom to take a strong stand against those who would prefer a theocracy not in Iran or Afghanistan, but in the United States."[49] Repeatedly, members of the biological establishment view ID as a threat to the integrity of biology as a science and as a threat to the very foundations of Western society, not simply a threat to the truthfulness of a single scientific theory. This reinforces the insight of Michael Denton above. Biologists recognize that if Darwinism is undermined, biology's status as a naturalistic science might be undermined with it. And this puts enormous pressure on the biological establishment to uphold the truth of Darwinian evolution even when the evidence does not support it. Hence, the creation of the grand narrative of Darwinian triumph.

Further reinforcing this point is that these calls to resist ID theory because of the threat it poses to the integrity of science and society seem only to come from biologists—those who have the biggest professional stake in the debate. Scientists from other fields tend to be more laconic in

---

47. Palevitz, "Intelligent Design Creationism," 1719.
48. Park, "It's Not about Evolution," 113.
49. Berman, "Intelligent Design Creationism," 646.

their treatment of the work of ID proponents. For example, in a review of ID theorist Michael Behe's *Darwin's Black Box*, biologist Neil Blackstone charges Behe with oversimplifying evolutionary theory, making implausible assumptions, committing errors in logic, ignoring relevant literature, and neglecting proper methodology.[50] Nicholas Matzke, reviewing Behe's later book *The Edge of Evolution*, writes, "It is clear that Behe is driven not by a truly scientific investigation, but instead by metaphysics."[51] And Jeffrey Shallit, reviewing ID theorist William Dembski's *No Free Lunch: Why Specified Complexity Cannot Be Purchased without Intelligence*, calls Dembski's book "a poorly written piece of propaganda and pseudomathematics."[52]

Now compare the review of *Darwin's Black Box* appearing in the journal *Integrative Physiological and Behavioral Science* by Stewart Wolf, a medical doctor. Wolf believes Behe's argument provides a strong inference to support the idea that the origin of life and the origin of complex adaptive systems were designed, and he concludes, "Behe's evidence leading to his theory must be taken seriously."[53] Likewise, physicist Leo Kadanoff, reviewing Dembski's *No Free Lunch*, considers Dembski's mode of argument to be fully within the traditions of science, even when invoking an implausible explanation when more plausible ways have been eliminated. Kadanoff makes clear that he does not accept the conclusions of ID theory ("Not yet anyway"), but he does think Dembski's work needs to be seriously engaged, not disparaged.[54] Are Wolf and Kadanoff really reading the same books as Blackstone and Shallit?! Another physicist, Robert Ehrlich, has compared ID theory to string theory, questioning whether string theory with its highly abstract and largely untestable predictions about the structure of the universe has any greater claim to being science than ID. Though he concludes that it does, the tone of his article engages ID in a serious and substantive way; it lacks the harsh ad hominem tone of the biologists.[55] Clearly, those whose professional identity is most threatened by ID engage in the sharpest rhetorical critique of it. Biologists know that if ID theory succeeds at undermining Darwinian

---

50. Blackstone, Review of *Darwin's Black Box*, 446.
51. Matzke, Review of *The Edge of Evolution*, 566.
52. Shallit, Review of *No Free Lunch*, 93.
53. Wolf, Review of *Darwin's Black Box*, 418.
54. Kadanoff, "Intelligent Design and Complexity Research," 453.
55. Ehrlich, "What Makes a Theory Testable."

evolution, biology may well lose its status as a naturalistic science. Medical doctors and physicists don't have so much at stake. Powerful guild interests therefore become the driving force in maintaining the grand narrative of Darwinian triumph.

Perhaps the biological establishment has forgotten the insight registered many years ago by one of the heroes of the evolutionary synthesis, the American paleontologist George Gaylord Simpson. In Simpson's seminal 1944 monograph *Tempo and Mode in Evolution*, a work credited with demonstrating that paleontological data is fully consistent with the gradual understanding of evolution being developed at the time by the population geneticists, Simpson unexpectedly waxes philosophical. He argues that rejecting all physical hypotheses to explain some phenomenon does not warrant proposing a metaphysical hypothesis as a part of scientific explanation. But he continues, "The metaphysical explanation might, indeed, be true, but no hypothesis for which there can be no rigidly objective test should be accepted as a conclusion or working principle in research."[56] Simpson surprisingly admits that metaphysical explanations of natural phenomena might turn out to be true. They just cannot be accepted within the bounds of science, which is defined as the search for naturalistic explanations. Simpson then doubles down on this point in an accompanying footnote:

> As a matter of personal philosophy, I do not here mean to endorse an entirely mechanistic or materialistic view of the life processes. I suspect that there is a great deal in the universe that never will be explained in such terms and much that may be inexplicable on a purely physical plane. But scientific history conclusively demonstrates that the progress of knowledge rigidly requires that no non-physical postulate ever be admitted in connection with the study of physical phenomena. We do not know what is and what is not explicable in physical terms, and the researcher who is seeking explanations must seek physical explanations only, or the two kinds can never be disentangled.[57]

Simpson seems to imply, whether he intends to or not, that since non-physical explanations may indeed be true, even though they cannot be considered by scientists, science cannot then claim to be the sole arbiter of truth.

56. Simpson, *Tempo and Mode in Evolution*, 76.
57. Ibid., 76 n. 13.

Defining a non-Darwinian understanding of evolution like that proposed by ID theorists as outside the bounds of science is not tantamount to rendering it false. It may not be science, but it may in fact be true. And science, by its decision to search only for naturalistic explanations, may not always lead us to truth. If Michael Denton is right that Darwinian evolution is the only truly *scientific* approach to evolution, then it is understandable why the biological establishment adheres to it so dogmatically. Now, biologists may be obliged to hold to a purely naturalistic explanation of evolution, but religion scholars and theologians need not be so obliged. When the question becomes, what is true? rather than, what is science?, Darwinian theory must be open to criticism.

Let me be clear that this book is not meant to be an argument for ID. Many books on the market do engage the positive arguments for this frequently disparaged theory, and I have little of substance to add to those arguments (I will consider ID in more detail in chapter 7). My position on ID is similar to the surprisingly sympathetic view taken by the atheist philosopher Thomas Nagel: "Even though writers like Michael Behe and Stephen Meyer are motivated at least in part by their religious beliefs, the empirical arguments they offer against the likelihood that the origin of life and its evolutionary history can be fully explained by physics and chemistry are of great interest in themselves."[58] Even if one is not drawn to the conclusions of ID thinkers, Nagel believes the arguments they raise against Darwinism should be taken seriously. "They do not deserve the scorn with which they are commonly met. It is manifestly unfair."[59] I agree. Once we realize that the triumph of Darwinian theory functions as a grand narrative, a story whose purpose it is to advance an ideological agenda more than it is to discern the truth of how evolution actually works, we can embrace the idea that criticisms of Darwinism are in fact intellectually and academically respectable and can open ourselves to the possibility of alternative explanations, of which ID is one. Religion scholars and theologians are thus freed from slavish capitulation to Darwinism to think more creatively about the positive role religious thought might play in accounting for the origin and diversity of life. But first, we will need to engage the historical development of evolutionary theory in order to grapple fully with its ambiguous and ideological nature. In

58. Nagel, *Mind and Cosmos*, 10.
59. Ibid.

doing this, we will be able to develop a narrative that stands counter to the ubiquitous grand narrative of Darwinian triumph.

Evolutionary Theory: A Counternarrative

How does the grand narrative of Darwinian triumph fundamentally misrepresent the historical development of evolutionary thinking? I will lay out here in brief a counternarrative to provide a map through the rest of this book, where this counternarrative will be developed in greater detail. In developing a counternarrative, it is interestingly not necessary to focus on the critics of Darwinian theory, for the scientific literature itself—literature designed to support the grand claims of the narrative of Darwinian triumph—deconstructs itself when read with a critical eye. This is what I will attempt to do.

We begin where the grand narrative begins, with Charles Darwin and the publication of *On the Origin of Species*. Who was Darwin, and what exactly did he accomplish? Because of the influence of the grand narrative, Darwin has become a larger-than-life figure, credited with having the greatest idea anyone has ever had, or with developing the most important theory in the history of science. I will retell the story of Darwin in a more nuanced way that attends not only to his particular genius but also to his very human flaws. Darwin's particular interest while he was sailing around the world on HMS *Beagle* was geology, not biology. And at least two of his geological theories were proven false in his day. Darwin had a penchant for developing theoretical speculations that went far beyond the evidence he had at hand. This seems to have been the case in his biological work as well.

It is a well-documented fact that Darwin was unable to convince the majority of his contemporaries that natural selection could be the primary mechanism of evolutionary change. This is mainly because he provided little evidence for it. Darwin commits a sort of bait and switch in the *Origin*. He provides copious amounts of empirical observation supporting the idea that evolution has occurred (what is sometimes called the fact of evolution). But this was not a particularly controversial idea in Darwin's day. Evolution was already in the air, as we will see. Darwin's real contribution was to hypothesize a naturalistic mechanism for evolution—natural selection (what he referred to as "my theory"). But lacking empirical evidence for natural selection, Darwin tried to make a strong empirical case for the fact of evolution and hoped his readers would simply buy natural selection along with it. Many readers didn't.

The exception most often cited is Thomas Henry Huxley, often referred to as Darwin's bulldog. Huxley did take on the mantle of Darwin's chief surrogate, engaging in public battles with Darwin's critics long after Darwin himself had grown tired of publicly defending his theory. But Huxley's support of Darwin had little to do with the truth of natural selection. He never fully accepted natural selection and himself espoused a saltational understanding of evolution quite at odds with Darwin's. What Huxley liked about the *Origin* was not the theory itself, but the way Darwin had gone about trying to answer the question of the origin of species. Huxley (who coined the term *agnostic*) was animated by a zeal to rid natural history of theological baggage and to set up the emerging field of biology as a purely naturalistic science. A harsh critic of the church establishment of his day, Huxley is in many respects the father of modern biology. *On the Origin of Species* was for Huxley the paradigmatic example of what good empirical, naturalistic science should look like, and it was on this basis that he championed it.

This in reality is what Darwin accomplished. The scientific significance of his work is dubious at best in terms of providing a likely answer to an empirical scientific question: why is life so extravagantly diverse? Darwin's real significance is not so much scientific as historical. He helped lead the way to the development of biology as a naturalistic science on a par with physics and chemistry.

Next, I will consider the eclipse and recovery of Darwinism in the early years of the twentieth century. With the rediscovery of Mendel's work on heredity in 1900, Darwinian theory lost even more support as early Mendelian geneticists like Hugo de Vries, William Bateson, and Thomas Hunt Morgan came to support saltational theories of evolution. One noteworthy exception was the German biologist August Weismann. Weismann was a sharp critic of the idea put forth by the French naturalist Jean Baptiste Lamarck in the early nineteenth century, called the inheritance of acquired characters. Lamarck thought that changes that occur in an organism by virtue of its interaction with its physical environment could be passed down to the next generation. Even Darwin accepted some aspects of Lamarck's theory. But Weismann did not. He believed that the germ cells of an organism (egg and sperm) were walled off from the rest of the organism in such a way that changes acquired by an organism during its lifetime could not possibly be passed on to the next generation. Weismann, therefore, became even more Darwinian than Darwin himself by upholding natural selection not as the main—but as the only—mechanism of evolution. But as we will see, Weismann admits that his support for natural selection was not based on direct evidence but, as he says, "on quite other grounds." In chapter 3, we will find that Weismann's "quite other grounds" for championing natural selection were primarily ideological, not empirical.

Despite Weismann's plea, non-Darwinian understandings of evolution continued to dominate in the early twentieth century. This began to change, however, with the development of population genetics. Ronald Fisher, J. B. S. Haldane, and Sewall Wright developed sophisticated mathematical models (so sophisticated that most biologists couldn't understand them!) designed to demonstrate how small inherited variations within populations of organisms could spread through the population and accumulate, leading to gradual evolutionary change. The grand narrative often treats Fisher, Haldane, and Wright as all of a piece, as if they were united in demonstrating how natural selection could work. But in reality, the three engaged in deep and bitter debates over fundamental issues that in most cases were never resolved. We will look at the importance of these deep divisions and observe other significant critiques of the work of the population geneticists. The significance of population genetics has much more to do with the way it provided a rigorous mathematical framework for evolutionary theory and helped evolution resemble physics, the queen of the sciences in the 1920s (the era of Einstein, relativity, and quantum theory). In no sense can the population geneticists be said to have proven that natural selection is the primary driving force of evolutionary change.

But the grand narrative does not necessarily posit the early development of population genetics as the era when Darwinian theory reigned triumphant. This is reserved for the period of the modern evolutionary synthesis, ushered in with the 1937 publication of Theodosius Dobzhansky's *Genetics and the Origin of Species*. Dobzhansky is credited with creating a synthesis between the work of naturalists and the work of the population geneticists and showing how the gradual Darwinian mechanism of natural selection, acting on the inherent variability of organisms, could explain the extraordinary array of species that fill the earth. But reading Dobzhansky's work, we find that he expressed far less confidence in his understanding of evolutionary mechanisms than the grand narrative would lead us to expect. And as a committed Russian Orthodox

Christian with an abiding interest in the highly teleological evolutionary theology of the Catholic priest turned paleontologist Pierre Teilhard de Chardin, Dobzhansky in many ways makes a poor poster child for the purely scientific Darwinian view of evolution he is credited with establishing. (Dobzhansky even called himself a creationist!)

Despite statements to the contrary, the period of the modern evolutionary synthesis did not end with natural selection established as the clear mechanism of evolution. Many of the seminal texts of the synthesis are far more ambiguous than the grand narrative portrays. The paleontologist George Gaylord Simpson felt compelled to introduce the non-Darwinian concept of quantum evolution to account for the evidence of rapid evolutionary change documented in the fossil record. Bernhard Rensch's *Evolution above the Species Level*, though it is credited with showing how natural selection can account for large-scale, macroevolutionary trends, is itself a highly ideological (and tendentious) work that has several glaring internal inconsistencies. And perhaps most important, the shapers of the synthesis were unable to convince one of the most respected geneticists of the first half of the twentieth century of the truth of Darwinian theory. Richard Goldschmidt always held to a saltational approach and became one of the chief antagonists and most substantive critics of the shapers of the synthesis. Even Dobzhansky and Sewall Wright were forced to admit that Goldschmidt's criticisms were substantive and worthy of consideration.

Whatever sense we make of the evolutionary synthesis of the 1940s, Darwinian theory seemed to get a boost from one of the great advances in science—the 1953 discovery of the double helical structure of the DNA molecule by Francis Crick and James Watson. This scientific milestone ushered in the era of molecular biology, a discipline destined to revolutionize the study of life. Early on, Crick pronounced what he termed the Central Dogma of molecular biology. The Central Dogma states that information can flow in only one direction, from DNA to its transcription into RNA to the synthesis of proteins. Information stored in proteins cannot, according to this dogma, move backward and influence DNA sequences. At the time he pronounced it, Crick admitted that the evidence for the Central Dogma was negligible, but that simply based on his experience it seemed to be the way things worked. It should not escape our notice that Crick was originally trained as a physicist and only turned to biological questions out of a desire to stamp out what he saw as the last vestiges of vitalism remaining in biology, a desire consistent with

Crick's strong atheistic commitments. More important perhaps is the way the Central Dogma supports a Darwinian view of evolution.

For evolution to occur, new kinds of proteins have to evolve to create new kinds of organic structures. Since protein synthesis is coded by DNA sequences, new DNA sequences have to emerge in order to give rise to new kinds of proteins. If no outside influence can reach back and alter DNA sequences in any intentional way, as the Central Dogma states, then the only way DNA sequences can change is through the process of random mutation, the random copying errors that occur during the process of DNA replication. Thus, evolution must be a nondirected, non-purpose-driven process. Variability in organisms is produced by random alterations occurring during DNA replication, and occasionally, those DNA changes just happen to lead to an organism with a variation rendering it more fit in its environment. This variation is then preserved and perpetuated by natural selection, leading to incremental evolutionary change. Over enormous spans of time, this process can account for Earth's entire evolutionary history. The Central Dogma of molecular biology and Darwinian evolutionary theory fit together hand in glove. But given that Crick said the direct evidence for the Central Dogma was negligible, how do we know it is true?

As early as the 1950s, evidence seeming to contradict the Central Dogma began to emerge in the work of the Nobel Prize-winning corn geneticist Barbara McClintock. This was furthered in the 1980s when John Cairns and his associates published the results of bacterial experiments that they believed demonstrated nonrandom mutation. James Shapiro has coined the term "natural genetic engineering" to describe the way he believes cells have the ability to monitor their environments and intentionally alter their DNA sequences in order to respond to environmental challenges—a process, if it occurs, in clear violation of Crick's Central Dogma.[60] If McClintock, Cairns, Shapiro, and others are right, evolution, contra Darwin, might be a highly intentional process. The controversy over directed mutation continues unresolved, and one cannot help but think that the controversy has much less to do with the workings of a particular biological mechanism, and much more to do with the larger philosophical (and even theological) implications embedded in the controversy. Debates about directed mutation are frequently couched in terms of orthodoxy versus heresy. As any religion scholar knows, these

---

60. Shapiro, *Evolution*, 43.

terms originated in early Christian doctrinal debates, not as a way of sorting out truth from error, but as a way of sorting out who would possess the power to decide what is truth and what is error. That a term such as "Darwinian orthodoxy" shows up so frequently in the scientific literature indicates strongly the ideological nature of these supposedly scientific debates. If natural genetic engineering does occur, it would exact a fatal blow to a strictly Darwinian view of evolution, which could be seen by some as undermining all of biology as a naturalistic science.

In light of this, it is interesting to note the way proponents of Darwinian evolution try to develop analogies to help the uninitiated understand how natural selection works. This began with Darwin himself, who famously compared natural selection to the actions of artificial selection exercised by plant and animal breeders. Think of how artificial selection has led to the dizzying array of dog breeds, for instance. Whether it is August Weismann's architect, Francisco Ayala's artist, François Jacob's tinkerer, or Ernst Mayr's automotive engineer, all these analogies are based on intentional activity. Architecture, painting, engineering, even tinkering, are all highly purposeful undertakings. Evolutionary biologists have simply failed to find an analogy that actually supports a nonteleological (nondirected) view of natural selection. Along with these analogies, we also find teleological language creeping into the scientific literature under the cover of peer-reviewed research, as when natural selection is described as "opportunistic," or is said to "sculpt" organisms into something that looks designed. If Darwinian proponents cannot talk about evolution without inadvertently introducing teleological ideas into their discussions of a supposedly nonteleological process, we may need to pay much more attention to the scientific evidence that seems to support a more teleological view. Evolution may turn out to be more intentional than we think.

This raises an important question: is the modern evolutionary synthesis that enshrined Darwinian ideas at the heart of evolutionary theory past its prime? A growing chorus of biologists is calling for a fundamental rethinking of evolutionary theory. In the early 1970s, the theory of punctuated equilibria developed by Niles Eldredge and Stephen Jay Gould attempted to bring about a qualitatively new understanding of evolutionary mechanisms, but the Darwinian establishment succeeded in absorbing this idea within the framework of Darwinism, a movement to which Eldredge and Gould largely capitulated. But others have not. Robert G. B. Reid went to his grave holding to a sharply critical view of

Darwinian theory, while Eva Jablonka and Marian Lamb have revived the ghost of Lamarck in their work on epigenetic inheritance and evolution. In 2008, sixteen biologists met in Altenberg, Austria, to lay the foundation for an extended evolutionary synthesis designed to address the evident weaknesses of Darwinian theory. And in any summary of contemporary challenges to Darwinian evolution we cannot forget the ID movement. Darwinism may still be the established theory, but cracks continue to widen in its foundation. Dogged adherence to it has more to do with its ideological importance than its scientific value. In the twenty-first century it is no longer necessary to be deceived by Darwin.

This is a brief outline of the alternative history I will develop in the following chapters. Along the way, I will pause to consider some of the more scientifically substantive criticisms of Darwinian theory. These will include discussions of convergent evolution and the evolution of the eye; and discussions of classic works on natural selection by John Endler and Graham Bell; along with assessment of the phenomena of industrial melanism, *polyploidy* (the sudden emergence of new species in hybrid plants), and the Cambrian explosion (the rapid appearance of animal body plans 530 million years ago); and finally will include an examination of the possibility of directed mutation.

Having revealed the story of Darwinian triumph to be a grand narrative that overrides a long, complex, and ambiguous history, I will conclude by considering some ways that religion scholars and theologians might approach the religion–science relationship differently. If it is not necessary to always capitulate to Darwinism, what positive contributions can religion scholars and theologians make to the question of life and its diversity? Are there good reasons not to abandon traditional religious notions of an intelligent creator? What would it mean for religion scholars and theologians to embrace James Shapiro's view that evolution very much remains a mystery? Can embracing mystery be understood, not as an abdication of the duty to advance scientific progress, but rather as a humble acceptance of the limits of human reason, as an embracing of objective reality, and as a fundamentally religious orientation to life? Questions like these will occupy us in the final chapter.

Those committed to the grand narrative of Darwinian triumph will not find the counternarrative outlined here very persuasive. I will be accused of generating a simplistic portrayal of modern evolutionary theory that ignores the empirical work of an army of evolutionary biologists that fills many library shelves' worth of peer-reviewed journals. Perhaps these critics will have a point, but I respectfully ask that readers suspend judgment until after considering the more detailed discussions that follow.

2

Reframing Darwin and Darwinism

GERTRUDE HIMMELFARB ONCE POSED a most insightful question about Charles Darwin's *On the Origin of Species*: "By what metamorphosis did a scientific treatise, largely devoted to such abstruse matters as anatomical variations among different breeds of pigeons, become a metaphysics, politics, and economics?"[1] While Himmelfarb certainly overstates the centrality of pigeons in Darwin's seminal work—there are many other organisms discussed—we should not ignore the larger issue to which she points. Darwin's famous work *is* a book devoted to the piling up of anecdotal observations about the most seemingly insignificant features of the most mundane of living organisms collected from correspondents living in the far-flung regions of the British Empire. It is difficult to understand how such a dense, arcane, and technical monograph came to revolutionize the entire culture of Western civilization in such a profound and wide-ranging way.

Given the argument of the preceding chapter, Himmelfarb's question may lose some of its mystery. In stark contrast to the status it enjoys today, *On the Origin of Species* did not have a profound impact on the majority of Darwin's contemporaries. They read the book, found it interesting, and accorded it polite respect, but did not see it as having sparked a major revolution in thinking about the origin and diversity of life—let alone in metaphysics, politics, and economics. The *Origin* did not achieve its canonical status until the period of the modern evolutionary synthesis of the 1940s and the concomitant development of a grand narrative of

---

1. Himmelfarb, *Darwin and the Darwinian Revolution*, ix.

Darwinian triumph designed to trumpet Darwin's theory in service to an emerging understanding of biology as a proper naturalistic science. Unfortunately, Charles Darwin's life and work are now so completely colored by this grand narrative that it is virtually impossible to engage this enigmatic Victorian figure in any kind of critically nuanced way. But we must try. Daniel Dennett may champion Darwin as having come up with the greatest idea anyone has ever had, but this assertion is nothing more than ideologically motivated hyperbole.[2] Like all humans, Darwin was a complex figure capable of both profound insight and demonstrable error. We need to develop a more nuanced picture of this unlikely progenitor of modern evolutionary theory.

What Did Darwin Do?

Darwin's scientific contribution cannot be appreciated apart from an understanding of Charles Darwin the person. Therefore, I will provide here a brief biographical outline that attends to the facts as we know them, and seeks to avoid the common tendency toward hagiography.

The Life of Charles Darwin

Charles Robert Darwin was born on February 12, 1809, in the village of Shrewsbury, England, to Robert Darwin, a prominent physician, and Susannah Wedgwood Darwin, heir to the Wedgwood china fortune. With two such prominent parents, Darwin's financial comfort was set from his birth. He was a member of the upper class of Victorian society and would never hold a paying job in his entire life.

This does not mean that Darwin did not endure hardship. His mother died when he was eight, leaving him to be raised by an authoritarian father and several doting older sisters. Shortly after his mother's death, Charles was sent by his father to a local boarding school that Charles hated. Though not a failure as a student, Charles did not impress his teachers as being particularly serious about academic learning. And as he grew, schoolwork became even less a priority, replaced by a desire for the sporting life. Darwin enjoyed riding horses, shooting, and quail hunting, activities he engaged in at every opportunity. Late in life, Darwin would recall with mortification the words of his father to him when

---

2. Dennett, *Darwin's Dangerous Idea*, 21.

he was young: "You care for nothing but shooting, dogs, and rat-catching, and you will be a disgrace to yourself and all your family."[3]

Eventually, Darwin's frustrated father sent him to the University of Edinburgh to study medicine—to follow in the footsteps of his father and grandfather. Darwin's older brother Erasmus was already at Edinburgh, a fact that made his transition to university life easier. But the younger Darwin found little of interest in the study of medicine, and he detested the requirement to observe surgery; the screams and blood turned his stomach. In a letter from his sister Susan, he learned of his father's disappointment over his penchant for picking and choosing what lectures to attend: his father warned him not to continue his indulgent ways.[4] One aspect of life at Edinburgh, however, did get Darwin's attention. Many of the professors had a side interest in natural history, and he began attending natural history lectures, which he found far more interesting than his medical studies. He soon became fond of collecting beetles, trying to identify new or unusual species. Darwin endured his time at Edinburgh while his brother Erasmus was there to accompany him, but when Erasmus graduated, the days of the younger Darwin at Edinburgh were numbered. Finally getting up the nerve to reveal to his father his revulsion at the prospect of medical career, Charles quickly found himself packed off to Cambridge to study for the ministry. Robert Darwin was determined to find for his wayward son a vocation worthy of the family's social status.

As at Edinburgh, so at Cambridge Darwin found natural history far more interesting than his formal ministry studies. He continued to collect beetles and accompanied faculty members on extended collecting excursions. He became especially close to the botanist John Henslow, with whom he spent many hours in the British countryside. Having completed his ministry studies with adequate, though not outstanding, grades, Charles Darwin began to plan a more extensive natural history expedition to the island of Tenerife. He had become obsessed with the tropics from reading the travelogues of Alexander von Humboldt, and he reports spending mornings in Cambridge gazing at the palm trees in the hothouse.[5] But before he could undertake this journey, an opportunity arose for a much greater adventure.

3. Barlow, ed., *Autobiography of Charles Darwin*, 28.

4. Letter from Susan Darwin to Charles Darwin dated March 27, 1826. In Burkhardt et al., eds., *Correspondence*, 1:37.

5. Letter from Charles Darwin to Caroline Darwin dated April 28, 1831. In Burkhardt et al., eds., *Correspondence*, 1:122.

The British admiralty was commissioning HMS *Beagle* for a three-year voyage to map the coast of South America under the leadership of Captain Robert Fitzroy. This would be Fitzroy's second adventure at sea, and he knew the rigors of such a trip. He had already experienced the loneliness of long months on the open ocean and feared for his mental health. Fitzroy's father had gone mad at sea and committed suicide, a fate that he was almost sure would befall him. So he sought a companion, someone of his own high social standing, who could engage him in intellectual conversation over meals and keep him grounded. If this person had some ability in natural history as well, all the better. So Fitzroy contacted John Henslow at Cambridge with the invitation to join the *Beagle* voyage. But Henslow was early in building his career at Cambridge, and his wife had recently given birth. Henslow regretfully turned down the offer but suggested Darwin be invited in his place. Though Darwin's father initially forbade him to accept, viewing such a trip as a waste of time and money, he later relented on the advice of his brother-in-law, Josiah Wedgwood, and Darwin sailed on December 27, 1831.

The planned three-year voyage turned into a nearly five-year-long ordeal during which the *Beagle* circumnavigated the globe while Darwin learned that he was prone to seasickness! But the years spent mapping the coast of South America afforded Darwin long stretches of time to live on land and explore his new surroundings. This he did with zeal, collecting all kinds of animal, insect, bird, and fossil specimens, which he shipped back to England to enhance the collections of British museums. His real interest, however, was geology.

While aboard the *Beagle*, Darwin read the Scottish geologist Charles Lyell's newly published *Principles of Geology*. Lyell had transformed geological thinking by replacing the popular doctrine of *catastrophism* with *uniformitarianism*. The doctrine of catastrophism taught that the earth's geological features had resulted from a series of catastrophic events similar to the global flood envisioned in the Bible. Uniformitarianism, by contrast, taught that the earth's geological features had resulted from the slow and gradual reshaping of the earth's surface by the same natural processes occurring today but acting over enormous spans of time—processes carried out by wind, water, volcanic eruptions, earthquakes, and so forth. Darwin was convinced by Lyell's theory and spent much of his time during the *Beagle* voyage looking for evidence to support it. The problem of life's mind-boggling diversity—or the species problem, as Darwin would later call it—does not appear to have been at the center of his attention during his five years at sea. The idea that Darwin was inspired to his biological views while observing finches in the Galápagos Islands appears to be more myth than reality. Darwin famously did not even label his finches by island, ignoring the potential significance of variations in beak size until they were pointed out to him by the ornithologist John Gould after his return to England.[6]

During the closing days of the voyage, Darwin fully expected to settle down to the quiet life of a country pastor. Becoming a famous scientist was far from his mind. But no sooner had the *Beagle* docked back in England in October of 1836 than Darwin found himself in great demand within the scientific community. The vast collection of specimens he had sent back over almost five years had attracted great attention, and his interest in finding evidence to support Lyell's geological theories made him a sought-after speaker. Darwin began to move through the rarefied air of the British scientific establishment in London, and he set about writing up geological papers along with his personal journal for publication. Darwin developed personal relationships with some of the most prominent scientists of his day, most important being Charles Lyell himself.

At some point, Darwin began to think about the relationship between the observations he made while on the *Beagle* and the idea of evolution, or what he called the transmutation of species. The geographical distribution of species around the world that Darwin had observed began to reinforce in his mind the very real likelihood that life had evolved over an enormous span of time. But Darwin well knew that evolution was not a new idea. His own grandfather, Erasmus Darwin, had put forth an evolutionary viewpoint in the late eighteenth century, as had several other notable thinkers. Most important for Darwin's developing thought was the work of the French naturalist Jean Baptiste Lamarck.

Lamarck believed not only that living organisms had changed over time, but that this change was driven in a particular direction. Life, according to Lamarck, had begun in a form both simple and imperfect, and had progressed over time to become more complex and closer to perfection. Lamarck attributed this natural tendency to organic change to what he called "the power of life," a force inherent in organisms that

---

6. The mythology surrounding Darwin's time in the Galápagos Islands has been explored in Sulloway, "Darwin and His Finches"; and more recently in Hennessey, "Mythologizing Darwin's Islands." Hennessey places the development of the legend of Darwin discovering evolution in the Galápagos in the context of the 1930s and attempts made to promote conservation programs in the islands.

propels them forward toward greater levels of perfection.[7] In addition, Lamarck hypothesized that evolutionary change in organisms could be shaped directly by the influence of the natural environment. This point came to be codified in the doctrine of "the inheritance of acquired characters," an idea that has become synonymous with Lamarckism, but which actually played a relatively minor role in Lamarck's own thought. Lamarck argued that as organisms interact with the environment during the course of their life, they undergo small changes. These small changes (acquired characters) are then passed down to the next generation. Over time, large-scale evolutionary change results. The textbook example of the inheritance of acquired characters is the long neck of the giraffe. Short-necked giraffes in the past would have strained their necks to graze on leaves a little higher in the trees. This would cause their necks to stretch slightly, a character they would pass down to their descendants. Over time, the giraffe's neck would get progressively longer leading to the ungainly long-necked giraffe of today.

Darwin read Lamarck's work and was deeply influenced by it. But whereas Lamarck understood his theory to represent a naturalistic understanding of evolution, his talk of "the power of life" was interpreted by many as a form of vitalism. The charge of vitalism was further reinforced by Lamarck's view that life had evolved in a particular direction, from simple to complex. It appears that Darwin was driven by the desire to develop an understanding of evolution that was not vitalistic in any way. He wanted a fully naturalistic explanation that did not view evolution as proceeding in any preconceived direction. The influence of Lyell's geology on Darwin's emerging biological thinking should not escape our notice. The idea that God had created each species in a unique act paralleled very closely the catastrophic view of geology. Darwin sought an explanation of evolutionary change that would parallel Lyell's uniformitarian view of geological change—that species evolve slowly and gradually over long periods of time, driven by the same purely naturalistic forces we see at work in the world today: variation, reproduction, and selection.

Lamarck's evolutionary theory was not the only influential pre-Darwinian contribution to the subject. In 1844, an anonymously authored book, *The Vestiges of the Natural History of Creation*, appeared in England with great fanfare. It laid out a grand evolutionary narrative in such a popular style that more than thirty thousand copies were eventually sold.

7. Burkhardt, *Spirit of System*, 151.

Vestiges provided little empirical evidence to support its evolutionary scenario, and the book was roundly ridiculed by the scientific establishment, including Darwin. But its popular appeal propelled evolutionary thinking to the forefront of British society, much to the chagrin of the ecclesiastical establishment. *Vestiges* was eventually revealed to be the work of the publisher Robert Chambers. Whereas Darwin treated Chambers's book with a level of contempt, he did recognize its importance in paving the way for his own work: "In my opinion it [*Vestiges*] has done excellent service in this country in calling attention to the subject, in removing prejudice, and in thus preparing the ground for the reception of analogous views."[8]

Darwin did not really have to convince people of the fact of evolution; evolutionary thinking was already prominent among members of the educated classes. He did, however, try with some success to provide a more thorough empirical basis for evolution than either Lamarck or Chambers had. But if Darwin was going to make a real contribution to science, he needed to posit a believable naturalistic mechanism for how evolution happens. In Darwin's own recounting, he stumbled upon a potential mechanism while reading Thomas Malthus's well-known *Essay on Population*. Malthus famously observed that human populations tend to increase geometrically while food supplies tend to increase only arithmetically—meaning that human populations will quickly outstrip their food supply, leading to mass starvation unless the population growth is limited in some other way. Darwin's flash of insight was to generalize Malthus's theory to populations of nonhuman organisms. The individuals of all species, Darwin suggested, would be locked in a struggle for scarce resources, but if some individuals happened to possess characteristics making them more successful in this struggle, they would be more likely to reproduce and pass on their more fit characteristics to the next generation. Over time, this struggle for existence would create organisms better adapted to their environment. But since environments constantly undergo change, what is adaptive in one context may not be in another, meaning that evolutionary change would never cease. This process of adaptation driven by natural selection acting over enormous spans of time would be able, Darwin believed, to account for the great diversity of species populating the earth.

Darwin would come to call this theory of an evolutionary mechanism "my theory." But Darwin was not alone in developing the concept

---

8. Darwin, *On the Origin of Species*, 6.

of natural selection. The English explorer and naturalist Alfred Russel Wallace came up with a virtually identical idea while living in the jungles of Southeast Asia. And like Darwin, Wallace also claimed to have been inspired by Malthus. Darwin had been in contact with Wallace over a period of years and knew of the work Wallace was doing among the indigenous people of the Malay Archipelago. But Darwin never imagined that Wallace would hit upon the same mechanism of evolutionary change as he had. Darwin was thus devastated when in 1858 he received from Wallace a copy of a paper on natural selection that Wallace hoped Darwin would help him publish. Instead, Darwin saw that the major contribution to science he hoped to make might well be usurped by Wallace, leaving Darwin in a quandary about how to respond to Wallace's request for publication assistance. With the help of Charles Lyell, it was decided that Darwin and Wallace's papers would be read together before the Royal Society. This allowed Darwin time to set about writing up his ideas in a more detailed form, a document that would appear in November 1859 as *On the Origin of Species*. With the publication of the *Origin*, Wallace's short paper fell into obscurity, and Darwin's fame as the face of evolutionary theory was assured. But was the *Origin* really such a revolutionary work?

Assessing *On the Origin of Species*

To accurately assess Darwin's biological work it will be helpful to first take note of two geological theories he proposed that were later proved to be in error. We learn much about how Darwin's mind worked from these two episodes, insights that will shed considerable light on his work in the *Origin*.

The first episode concerns a journey Darwin made in 1838 to the Scottish Highlands to examine mysterious markings known as the parallel roads of Glen Roy. These strange parallel markings on the land had engendered a considerable amount of scientific speculation as to their origin. What natural force could make such perfectly parallel lines? Darwin concluded that the area of Glen Roy had once been below sea level but had gradually been raised to its present location through the process of uplift. The parallel markings he attributed to the actions of the sea. Darwin presented his conclusions in a paper read before the Royal Society in 1839. But not long after, two geologists proposed an alternative theory—that the markings resulted from the action of glaciers scouring the land during the last ice age. This explanation was far more credible and quickly won the day. Reflecting on this episode in his autobiography, Darwin commented:

> During these two years I took several short excursions as a relaxation, and one longer one to the parallel roads of Glen Roy, an account of which was published in the *Philosophical Transactions*. This paper was a great failure, and I am ashamed of it. Having been deeply impressed with what I had seen of the elevation of the land in S. America, I attributed the parallel lines to the action of the sea; but I had to give up this view when Agassiz propounded his glacier-lake theory. Because no other explanation was possible under our then state of knowledge, I argued in favour of sea-action; and my error has been a good lesson to me never to trust in science to the principle of exclusion.[9]

Darwin could have avoided this error and the resulting embarrassment it engendered. In a letter to Charles Lyell explaining his theory, Darwin wrote, "I have fully convinced myself (after some doubting at first) that the shelves are sea-beaches, although I could not find the trace of a shell; and I think I can explain away most, if not all, the difficulties."[10] The lack of any trace of marine organisms should have tipped Darwin off that Glen Roy had not been under the sea in any geologically recent time. But when Darwin developed what he felt was a compelling idea, he doggedly held to it even when faced with a lack of clear evidence. And he became a master at trying to "explain away" difficulties that his theories often ran into. This characteristic is evident also in the *Origin*. At least in the Glen Roy episode, Darwin did relent and admit he was wrong. This was not the case with his work on coral reefs, however.

Based on observations Darwin made while aboard the *Beagle*, he developed a theory to explain the formation of coral reefs, which he published in 1842. Darwin believed that coral reefs resulted from the subsidence of the land. As a land mass subsided into the ocean, a fringe reef would form around it. With further subsidence a fringe reef would develop into a full barrier reef. Eventually, as the land continued to subside below the level of the ocean, a coral atoll would be left behind. As elegant as his theory was, Darwin really could not point to much in the way of

---

9. Barlow, ed., *Autobiography of Charles Darwin*, 84.

10. Letter from Darwin to Lyell dated Aug. 9, 1838. In Burkhardt et al., eds., *Correspondence*, 2:96.

evidence to prove it. Gertrude Himmelfarb comments, "The only 'proof' he could point to was the fact that there seemed to be no other way of accounting for the phenomenon."[11] Despite the lack of direct evidence, however, Darwin's theory was carefully argued and even Lyell came to accept it.

But further work by geologists began to accumulate evidence that some coral reefs originated as a result of land under the sea rising upward with the accumulation of small organic organisms, eventually reaching near the sea surface. Darwin's theory of subsidence might explain the origin of some coral reefs, but it could not be generalized as the only explanation. Darwin did come to accept that a theory of land elevation might explain some isolated examples of coral reefs and coral islands, but he still felt his theory of subsidence held in the vast majority of cases. In 1881, just a year before his death, Darwin wrote to Alexander Agassiz, a proponent of the "bottom-up" approach to coral reefs:

> Pray forgive me for troubling you at such length, but it has occurred to me that you might be disposed to give, after your wide experience, your judgment. If I am wrong, the sooner I am knocked on the head and annihilated so much the better. It still seems to me a marvelous thing that there should not have been much and long-continued subsidence in the beds of the great oceans."[12]

Darwin seems never to have fully accepted that his approach to the creation of coral reefs might be wrong.

We learn from Darwin's geological work that he was a keen observer of the world around him. But more than that, he had a constant need to make sense of his observations by proposing grand theories to explain them. His theories often went well beyond the actual evidence he had at hand, but once he proposed a grand explanatory theory, he was reluctant to give it up even in the face of accumulating contrary evidence. As Walter Cannon has commented, "Two of Darwin's chief virtues were ones not usually praised in a scientist: jumping to conclusions which go well beyond the evidence available; and maintaining his faith in his position regardless of the valid arguments that could be brought against it."[13]

11. Himmelfarb, *Darwin and the Darwinian Revolution*, 102.

12. Letter from Darwin to Agassiz dated May 5, 1881. In Darwin, ed., *More Letters of Charles Darwin*, 198.

13. Cannon, Foreword to *Charles Darwin: The Years of Controversy*, by Peter J. Vorzimmer, xv.

It is not clear that these are scientific virtues, but it is clear that these characteristics extended also to Darwin's attempt to account for the origin of species.

Even as staunch a supporter of Darwinian evolution as Ernst Mayr was forced to admit:

> One must grant Darwin's opponents the validity of two of their objections. First, Darwin produced embarrassingly little concrete evidence to back up some of his most important claims. This includes the change of one species into another in succeeding geological strata, or the production of new structures and taxonomic types by natural selection.[14]

This is a startling admission coming from as strong a Darwinian apologist as Mayr. But as we will see, most of Darwin's contemporaries agreed. The *Origin* does provide evidence to demonstrate that an evolutionary process is likely to have occurred, but Darwin's unique contribution, the theory of natural selection as the primary mechanism of evolution, stands as simply a postulate with little empirical support. This fact has led Barry Gale to give his book on Darwin's work the surprising title *Evolution without Evidence*. Michael Denton reports that Darwin freely admitted to a lack of hard, empirical evidence to explain the major evolutionary transformations he envisioned[15]—a point supported by a letter Darwin wrote to the Harvard botanist Asa Gray, in which he included a short abstract of his theory. Darwin wrote, "This sketch is *most* imperfect; but in so short a space I cannot make it better. Your imagination must fill up many wide blanks. Without some reflection, it will appear all rubbish; perhaps it will appear so after reflection."[16] Darwin was acutely aware that he was developing a very controversial theory—he famously shared with his botanist friend Joseph Dalton Hooker that confessing to his belief in evolution felt a little like confessing a murder!—and despite his own belief in his theory, he knew he did not have the kind of evidence that would convince skeptical readers. This is almost certainly why Darwin waited so long to publish. It was only his fear of losing priority when Wallace's manuscript arrived in the mail that forced Darwin to go public.

---

14. Mayr, "Evolution and God," 285.

15. Denton, *Evolution: A Theory in Crisis*, 117.

16. Letter from Darwin to Gray dated Sept. 5, 1857. In Burkhardt et al., eds., *Correspondence*, 6:449.

But go public he did in November 1859, only to face considerable backlash from critics. Darwin felt these criticisms acutely and ended up revising the *Origin* five times over the next two decades in order to answer his critics, but with limited success. The work became longer with each succeeding edition, but lacking any hard evidence, "the story of Darwin's handling of his theory of natural selection after 1859 is one of documented qualification and nagging doubt."[17] Rather than showing his critics to be wrong, Darwin ended up capitulating to them. The sixth and final edition of the *Origin* even contained an entirely new section on the "Effects of the increased Use and Disuse of Parts, as controlled by Natural Selection," in capitulation to the supporters of Lamarckism. Darwin here argues that atrophy attending the disuse of some body parts is heritable and accounts for the evolution of flightless birds and blind mammals. This is a classic case of Lamarckian inheritance of acquired characters.

Darwin's capitulation to critics led to inconsistencies in later editions of the *Origin*. Loren Eiseley has observed, "A close examination of the last edition of the *Origin* reveals that in attempting on scattered pages to meet the objections being launched against his theory the much-labored volume had become contradictory."[18] In one place, Darwin argued that organisms higher in the scale of nature evolve more quickly than those lower on the scale. But in another, he indicated that evolution was likely to have occurred more rapidly in its early stages when the earth would have been populated only by lower forms of life. Eiseley comments, "The last repairs to the *Origin* reveal . . . how very shaky Darwin's theoretical structure had become. His gracious ability to compromise had produced some striking inconsistencies."[19]

In a striking passage in the *Descent of Man*, Darwin's later treatise on human evolution, Darwin pleaded guilty to placing too much emphasis on natural selection as the primary mechanism of evolution:

> Thus a large yet undefined extension may safely be given to the direct and indirect results of natural selection; but I now admit, after reading the essay by Nageli on plants, and the remarks by various authors with respect to animals . . . that in the earlier editions of my 'Origin of Species' I perhaps attributed too much to the action of natural selection or the survival of the fittest. I have altered the fifth edition of the 'Origin' so as to confine my

---

17. Vorzimmer, *Charles Darwin*, 20.
18. Eiseley, *Darwin's Century*, 242.
19. Ibid.

> remarks to adaptive changes of structure; but I am convinced, from the light gained during even the last few years, that very many structures which now appear to us useless, will hereafter be proved to be useful, and will therefore come within the range of natural selection. Nevertheless, I did not formerly consider sufficiently the existence of structures, which, as far as we can at present judge, are neither beneficial nor injurious; and this I believe to be one of the greatest oversights as yet detected in my work.[20]

Following this passage, Darwin reminds his readers that his purpose in the *Origin* was twofold: to provide evidence against the idea of the special creation of organisms, and to argue for natural selection as the primary mechanism of evolutionary change. He then comments: "if I have erred in giving natural selection great power, which I am very far from admitting, or in having exaggerated its power, which is in itself probable, I have at least, as I hope, done good service in aiding to overthrow the dogma of separate creations."[21] Later in life, Darwin was obviously more confident in his contributions to the establishment of the fact of evolution than in his contributions to the question of evolutionary mechanisms.

These passages beautifully demonstrate much of what is said above. Darwin simply ignored evidence that did not fit with his theory, but remained convinced that he would be vindicated by the future discovery of new evidence. He made the same argument in the *Origin* in reference to the lack of paleontological evidence for his theory. Darwin simply hoped that fossil discoveries in the future would eventually prove him right. It never seemed to occur to him to abandon his theory in light of negative evidence. Peter Vorzimmer comments, "Though he lived to see many of his beliefs undermined, and at the end was no longer able to demonstrate the supremacy of his theory of natural selection, Darwin died consistent in at least one thing—his faith in the theory that bears his name."[22]

What drove Darwin to hold so tenaciously to a theory that even he admitted was not based on incontrovertible evidence? We can only speculate. According to Barry Gale:

> Darwin undoubtedly wanted to make a major contribution to the corpus of scientific knowledge. I do not think that it mattered to Darwin, especially as a young man, where he achieved

---

20. Darwin, *Descent of Man*, 441.
21. Ibid., 442.
22. Vorzimmer, *Charles Darwin*, 267.

recognition: in geology, zoology, or evolutionary theory. The important thing was that he achieve some noteworthy place, some 'fair place among scientific men,' as he put it.[23]

Darwin never held an academic appointment like other great figures in science. He never even held a paying job. He may have labored under the influence of his father's prediction that he would be a disgrace to the family. Darwin was clearly driven by the desire to make a mark, leaving him unable to give up faith in his signature theory, even if he struggled to convince others of its truth. Interestingly, after admitting that Darwin had provided "embarrassingly little concrete evidence" to support his theory, Ernst Mayr ascribes the reluctance of Darwin's contemporaries to accept natural selection to their inability to fully understand the radical nature of what Darwin was proposing.[24] But maybe they understood more than Mayr is willing to admit. Maybe they just failed to see the evidence. Exactly how did Darwin's close scientific colleagues react to the *Origin*?

Few scientists of the time were closer to Darwin than Charles Lyell. Given Darwin's strong support for Lyell's uniformitarian geology their close relationship does not come as a surprise. Yet Lyell was unable to provide the same unqualified support for Darwin's work that Darwin had afforded Lyell's. Shortly after the publication of the *Origin*, Lyell wrote to Darwin:

> I have just finished your volume and right glad I am that I did my best with Hooker to persuade you to publish it without waiting for a time which probably could never have arrived, though you lived till the age of a hundred, when you had prepared all your facts on which you ground so many grand generalizations.[25]

Lyell clearly noticed that the *Origin* was lacking in the evidence necessary to support its grand generalizations, though he seems to have thought this was because Darwin had rushed his treatise into publication. This idea is reinforced when later in the letter Lyell encourages Darwin: "I mean that, when, as I fully expect, a new edition is soon called for, you may here and there insert an actual case to relieve the vast number of abstract propositions." Lyell failed to read the *Origin* as a convincing demonstration of natural selection because the anticipated evidentiary

---

23. Gale, *Evolution without Evidence*, 146.

24. Mayr, "Evolution and God," 286.

25. Letter from Lyell to Darwin dated Oct. 3, 1859. In Burkhardt et al., eds., *Correspondence*, 7:339.

basis was simply not there. Lyell seems to have thought that Darwin must have had the evidence at hand and would be able to include it in future editions. But Lyell would be disappointed. Revised editions of the *Origin* became less convincing, not more. To be sure, Lyell would call Darwin's work "a splendid case of close reasoning" and showed that he was prepared "to take your statements of facts for granted." Despite his clear respect for Darwin's work, Lyell never bought into the full naturalistic and nonteleological implications of Darwin's theory, holding instead to the idea that "the whole course of nature may be the material embodiment of a preconcerted arrangement," and "the perpetual adaptation of the organic world to new conditions leaves the argument in favour of design and therefore a designer, as valid as ever."[26]

The nineteenth-century political philosopher John Stuart Mill also read the *Origin* and became a strong advocate, calling the *Origin* an "unimpeachable example of a legitimate hypothesis." But he further stated, "Mr. Darwin has never pretended that his doctrine was proved,"[27] and in another place wrote, "though he cannot be said to have proved the truth of his doctrine, he does seem to have proved that it *may* be true, which I take to be as great a triumph as knowledge & ingenuity could possibly achieve on such a question."[28] Mill was obviously a fan, but he clearly realized that the truth value of the *Origin* remained an open question.

Similarly, Darwin's strongest American supporter, Harvard botanist Asa Gray, recognized the *Origin* as an outstanding example of detailed empirical science and thought it deserved a wide reading. But Gray could never put his own Christian theological commitments aside and read the *Origin* as Darwin intended it to be read. Gray rather viewed the *Origin* as having brought teleological thinking back into natural history, a point that greatly frustrated Darwin, who spilled much ink trying to refute Gray's teleological conception of natural selection.[29] Overall, Darwin's close friends and colleagues found the *Origin* an important and interesting work deserving of a wide reading, but in general they were not persuaded that Darwin had come close to definitively solving the problem of the origin of species. Even Darwin's own brother, Erasmus, would write about the *Origin*, "In fact, the a priori reasoning is so entirely satisfactory

---

26. Lyell, *Antiquity of Man*, 393.
27. Mill, cited in Browne, *Charles Darwin*, 186.
28. Ibid.
29. Mayr, *Toward a New Philosophy of Biology*, 245.

to me that if the facts won't fit in, why so much the worse for the facts is my feeling."[30] But what about Darwin's most vociferous and aggressive supporter, his bulldog Thomas Henry Huxley?

Huxley is a key figure in the grand narrative of evolutionary biology, becoming known as Darwin's bulldog for his supposed dogged public defense of Darwin's theory, even to the point of ridiculing Bishop Samuel Wilberforce in a public debate in which Huxley informed the bishop he would rather have an ape for a grandfather than a pompous member of the ecclesiastical establishment, at which point women were said to have fainted. If Huxley ever did debate Wilberforce in public, this story has clearly been inflated to celebrate Huxley's role as Darwin's most aggressive defender. In the words of contemporary evolutionary psychologist David Barash, "By the end of the 19th century, Thomas Huxley was perhaps the most famous living biologist, renowned in the English-speaking world as 'Darwin's bulldog' for his fierce and determined defense of natural selection."[31] But did Huxley support Darwin because he thought he was right? The story is more complicated.

Huxley once described the *Origin* as "a mass of facts crushed and pounded into shape, rather than held together by an obvious logical bond."[32] It is hard to read this as a compliment! How did Huxley really feel about Darwin's work? It is well known that Huxley held to a saltational (discontinuous) view of evolution and could not agree, for most of his life anyway, with Darwin's view of gradual evolutionary change. Upon reading the *Origin*, Huxley wrote to Darwin, "you have loaded yourself with an unnecessary difficulty in adopting 'natura non facit saltum' so unreservedly. I believe she does make *small* jumps."[33] As far as natural selection is concerned, Huxley wrote in a review, "It is not absolutely proven that a group of animals, having all the characters exhibited by species in nature, has ever been originated by selection, whether artificial or natural."[34] And two years before his death, Huxley could still write:

---

30. Letter from Erasmus Darwin to Charles Darwin dated Nov. 23, 1859. In Burkhardt et al., eds., *Correspondence*, 7:390. Darwin was apparently so taken with his brother's unusual praise of his work that he repeated it in a letter sent to Charles Lyell the next day. See letter from Darwin to Lyell dated Nov. 24, 1859 in ibid., 394.

31. Barash, *Natural Selections*, 101.

32. Huxley, *Darwiniana*, 25.

33. Letter from Huxley to Darwin dated Nov. 23, 1859. In Burkhardt et al., eds., *Correspondence*, 7:391.

34. Huxley, *Darwiniana*, 74.

> Those who take the trouble to read the first two essays, published in 1859 and 1860, will, I think, do me the justice to admit that my zeal to secure fair play for Mr. Darwin, did not drive me into the position of a mere advocate; and that; while doing justice to the greatness of the argument I did not fail to indicate its weak points.[35]

Huxley makes it clear that despite his aggressive advocacy of Darwin's right to have his ideas aired, Huxley did not want to be remembered as one who simply accepted the truth of Darwin's theory in total.

Over time, Huxley did move closer to Darwin's understanding of evolution. The discovery of the now famous archaeopteryx fossils in the 1860s representing an apparent link between reptiles and birds helped convince Huxley that evolution could occur via small, gradual steps rather than by saltations. He also came to accept the idea of progressive development after having argued strongly against it. Yet, according to Sherrie Lyons, Huxley "remained skeptical his entire life of the power of natural selection to create new species."[36] As late as 1893 Huxley argued that natural selection could not be accepted as the primary mechanism of evolutionary change until selective breeding experiments succeeded in producing varieties of organisms infertile with each other—or clearly distinct species.[37] Huxley apparently never came to accept Darwin's most important contribution to evolutionary thought. What, then, was the basis for Huxley's aggressive public advocacy of Darwin's work?

Huxley had a much larger agenda than simply trying to account for the origin of species. As a strong critic of ecclesiastical authority, Huxley (who coined the term *agnostic*) desired to free natural history from the clutches of the church and render it a naturalistic science. *On the Origin of Species*, for all its faults, was for Huxley exemplary in its attention to detailed empirical observation and its attempt to account for the origin of species by positing purely undirected, naturalistic mechanisms. Huxley was less concerned with the rightness or wrongness of natural selection and more concerned with the naturalistic implications of what Darwin had proposed. In Thomas Lessl's words:

> As a scientific practitioner, Darwin symbolized something of greater importance than the mechanism of evolution he had

---

35. Ibid., v–vi.
36. Lyons, *Thomas Henry Huxley*, 231.
37. Lessl, *Rhetorical Darwinism*, 210.

proposed. His effort to account for life's origins and development on strictly naturalistic grounds signaled the universal reach that science was destined to achieve once free to follow its own course.[38]

This reveals David Barash's error above when he claims Huxley's renown was based on his "fierce and determined defense of natural selection." Barash's Huxley is the false creation of the grand narrative of Darwinian triumph, not the real Huxley who emerges from an assessment of the actual historical sources. Huxley manifestly was not a fierce and determined defender of natural selection but only of Darwin's method of trying to answer the species question. Huxley had written to Darwin, "Nothing I think can be better than the tone of the book."[39] The tone, more than the content, was Huxley's main concern.

Darwin largely failed to convince his closest friends and colleagues of the truth of natural selection. Despite the great respect they had for Darwin, they recognized that he had largely failed to provide evidence for his theory. The best they could do is stand with Mill in stating that Darwin proved that his theory *may be* true, but he certainly had not proven that it *was* true. And these were Darwin's supporters. What about his critics? What did they think of the *Origin*? The Cambridge geologist Adam Sedgwick, with whom Darwin had studied, never accepted Darwin's theory and wrote to him that parts of the *Origin* made him laugh until his sides hurt while other parts he read with absolute sorrow because "I think them utterly false and grievously mischievous."[40] Darwin indeed had many sharp critics, though space will not permit a comprehensive consideration of their many criticisms. I will focus instead on just one, the criticism of the Catholic layman St. George Jackson Mivart, whose criticism Darwin himself considered one of the most formidable ever raised. Mivart did indeed raise an important question about natural selection that appears never to have been adequately resolved.

---

38. Ibid., 172.

39. Letter from Huxley to Darwin dated Nov. 23, 1859. In Burkhardt et al., eds., *Correspondence*, 7:390.

40. Letter from Sedgwick to Darwin dated Nov. 24, 1859. Ibid., 396. Darwin responded to Sedgwick two days later that he expected Sedgwick's negative reaction to his work, but was not aware that he harbored any mischievous motivations when writing the *Origin*. See letter from Darwin to Sedgwick dated Nov. 26, 1859. Ibid., 403.

Mivart and Evolutionary Convergence

St. George Jackson Mivart was an acquaintance of Darwin and respected his work. But he felt that natural selection could not be the whole story of how evolution occurs. In 1871 he published *On the Genesis of Species*, a title specifically intended to parallel Darwin's *On the Origin of Species*. Mivart's work was an extended critical engagement with the *Origin* in which Mivart attempted to show why natural selection could not fully account for the evolution of life, but needed to be "supplemented by the action of some other natural law or laws as yet undiscovered."[41] These undiscovered laws Mivart described as likely constituting "special powers and tendencies existing in each organism."[42] Further, he argued that "such an internal power is a great, perhaps the main, determining agent."[43] Mivart then takes the expected leap by connecting this internal power with the action of some sort of intelligence, though carefully adding that this intelligence is "evidently not altogether such as ours, or else has other ends in view than those most obvious to us."[44] Mivart admits that this view of evolution is consistent with theistic conceptions, though he points out that consistency with theism should not be used as an argument in favor of his own view. Mivart believed that his idea that evolution was driven by "an internal innate force" could be proven by the evidence. What evidence did he offer?

There are several lines of argument in Mivart's book, but the most significant is his recognition of the phenomenon commonly known today as convergent evolution (or convergence). Convergence describes the well-documented existence of organisms that though they do not share a close evolutionary history, nevertheless evolve into very similar forms. Compare, for example, the mammals of North America with those of Australia. North American and Australian mammals are not closely related in any evolutionary sense, evolving on entirely separate continents and developing some stark differences: North American mammals are placentals whereas Australian mammals are marsupials. Yet, North American and Australian mammals nevertheless have converged onto very similar forms. One will find on both continents a small cat (the ocelot in North America), an anteater, a mouse-like rodent, a flying squirrel,

41. Mivart, *On the Genesis of Species*, 17.
42. Ibid., 33.
43. Ibid., 242.
44. Ibid., 253.

and a wolf-like creature as the top predator. Evolutionary convergence is ubiquitous in nature, and more examples will be given below. But on what grounds did Mivart believe that the reality of convergence was inconsistent with the expectations laid out by Darwin's theory?

Mivart wrote:

> It would be easy to produce a multitude of such instances of similar ends being attained by dissimilar means, and it is here contended that by 'the action of Natural Selection' *only* it is so improbable as to be practically impossible for two exactly similar structures to have ever been independently developed. It is so because the number of possible variations is indefinitely great, and it is therefore an indefinitely great number to one against a similar series of variations occurring and being similarly preserved in any two independent instances.[45]

A Darwinian view of evolution centers on the idea that the variations on which natural selection acts are randomly generated. One would therefore expect organisms evolving on different continents under different environmental conditions to diverge into starkly different forms rather than to converge onto similar forms. For organisms to converge, similar kinds of variations would have to be produced repeatedly in populations living in widely divergent environments, and for Mivart it was impossible to understand how a random process could so often produce similar evolutionary outcomes. Even Ernst Mayr comments on the low probability of the same or similar mutations occurring in two populations in the same sequence, though he ignores the significance of this.[46] For Mivart the reality of convergence was more easily squared with the idea that some internal force drove evolution in particular directions. Evolution must be constrained in some way to continually reproduce in different environmental contexts a limited number of organismic forms. Convergent evolution continues to be a source of wonder today and stands at the center of work being done by the Cambridge paleontologist Simon Conway Morris.

As we will learn more in chapter 5, Darwin was greatly troubled by a phenomenon encountered in the fossil record known today as the Cambrian explosion. This refers to the sudden (geologically speaking) appearance in the fossil record of many new animal body plans during

---

45. Ibid., 79.
46. Mayr, *Animal Species and Evolution*, 527.

the Cambrian period, some 530 million years ago. Darwin's theory does not allow for the sudden appearance of such high-level taxonomic categories (new phyla); new phyla could evolve only as the result of a long, gradual series of evolutionary divergences at lower taxonomic levels, leading up to the emergence of new phyla. Darwin knew that the fossil record of his day ran counter to this gradualist understanding, but he hoped that future fossil discoveries would reveal a long process of pre-Cambrian evolution leading up to what only *appeared* to be the sudden emergence of new animal body plans in the Cambrian era. As we will see later, future discoveries have not vindicated Darwin on this point. But post-Darwinian fossil discoveries *have* greatly illuminated the phenomenon of the Cambrian explosion.

Early in the twentieth century, a large cache of Cambrian fossils exquisitely preserving strange animal forms were discovered high up in the Canadian Rockies in a formation known as the Burgess Shale. The Burgess Shale fossils have more recently been supplemented by Cambrian fossil finds in China and Greenland. Simon Conway Morris came to prominence in paleontology when as a Cambridge graduate student he had the opportunity to work on the Burgess Shale fossils under the direction of Harry Whittington. This work eventually led to the book *Crucible of Creation* (1998), in which Conway Morris lays out his interpretation of the meaning of the Burgess Shale fossils. But Conway Morris was not the first paleontologist to publish a comprehensive volume on the Burgess Shale.

Nine years before, the noted Harvard paleontologist Stephen Jay Gould had published *Wonderful Life: The Burgess Shale and the Nature of History*, a title meant to conjure up the famous 1940s Frank Capra Christmas movie *It's a Wonderful Life* starring Jimmy Stewart. In the movie, George Bailey (played by Stewart) learns that life is a contingent process. One small detail from the past turning out differently would have had enormous implications for Bailey's future life—implications that would also have a major impact on all the citizens of the fictional town of Bedford Falls. With this as an analogy, Gould interprets the Burgess Shale fossils as telling a similar story of contingency in the history of life. The accidental extinction of one species may send ripple effects down through the ages, affecting the whole evolutionary process in unpredictable ways. According to Gould, even if we replay the tape of animal evolution a million times from its beginnings in the Burgess Shale, "I doubt that anything like *Homo sapiens* would ever evolve again. It is, indeed, a wonderful life."⁴⁷ Since evolution is contingent, it is fundamentally unpredictable. Any new evolutionary process would produce new and unexpected results.

But Conway Morris disagrees. When he looks at the Burgess Shale fossils and life in general, he sees a story of evolutionary constraint and convergence, not contingency and unpredictability. Evolutionary scenarios playing out in different contexts do not seem to lead to new and unexpected results as Gould argues; rather we see similar forms emerging over and over again as if evolution is constrained by some force or law to repeatedly produce a limited number of similar forms. Like Mivart before him, Conway Morris sees convergence everywhere in the broad landscape of life, and also like Mivart he recognizes that the reality of convergent evolution constitutes a powerful challenge to the Darwinian view of evolution.

Where does Conway Morris see convergence? He sees it in the human-like cognitive sophistication and playfulness of the octopus.[48] He sees it in the way three highly distinctive web constructions evolved independently more than once among species of orb spiders in Hawaii.[49] He sees it in the ubiquity of bioluminescence, a phenomenon so convergent we can be confident that "on any earthlike planet this strangely beautiful light will be flickering from ocean abysses to fungi in rainforests."[50] He sees it in the similar signal designs that independently evolved in bats facing similar environmental challenges.[51] He sees it in the way gliding has evolved independently at least thirty times[52] or the way air-breathing in fish has independently evolved anywhere from thirty-eight to sixty-seven times.[53] He sees it even in the intelligence of a plasmodium.[54] Conway Morris sees evidence of convergence literally everywhere. Evolution seems to exhibit lawlike features that sharply constrain the possibilities of what can evolve:

47. Gould, *Wonderful Life*, 289.
48. Conway Morris, *Runes of Evolution*, 19.
49. Ibid., 88.
50. Ibid., 111.
51. Ibid., 138.
52. Ibid., 189.
53. Ibid., 222.
54. Ibid., 237.

> Wherever one looks it seems as if evolution has very little choice. If, then, we can discover the details of the metaphysical "map" across which life must navigate to the very few available solutions, then are we not on the threshold of a predictive biology? This view of life cuts cleanly across one of the central areas of neo-Darwinian thinking, an area that insists on the randomness of evolution and the unpredictability of the outcomes. That might explain why the adjectives associated with descriptions of evolutionary convergence are so often exclamatory; "remarkable," "surprising," "astonishing," "uncanny," and "stunning" are all routinely employed.[55]

Perhaps the emotional impact of convergence is what prevents most biologists from commenting on it. Conway Morris observes that though naturalists get excited over the great diversity of life while molecular biologists are awed by the great range of molecules and the sophistication of their biochemistry, the constraints on evolution and the inevitability of evolutionary convergence receive far less attention. This is because "if evolution is in some sense channeled, then this reopens the controversial prospect of a teleology; that is, the process is underpinned by a purpose."[56] And the prospect of teleology is one the biological establishment simply cannot abide. According to Conway Morris, the best way to enrage an evolutionary biologist is to sidle up and suggest that evolution has a remarkable directionality. "If you are lucky," Conway Morris quips, "all you'll need is a clean handkerchief to dab the spots of spittle, but sometimes the response is closer to foaming."[57] The evidence of evolutionary convergence is hard to ignore, and no less so is its challenge to the sufficiency of a Darwinian view of the evolutionary process. Does this mean Conway Morris believes some spiritual force channels evolution into particular repeating directions? Not necessarily.

Conway Morris is no creationist, and he has characterized ID as a form of "theological mischief." Whatever constrains evolution may well be some sort of physical law. Yet for Conway Morris, the complexity and beauty of convergence—what he calls Life's Solution—"can never cease to astound. None of it presupposes, let alone proves, the existence of God, but all is congruent. For some it will remain as the pointless activity of the

---

55. Ibid., 21.
56. Conway Morris, "Evolution," 8.
57. Conway Morris, *Runes of Evolution*, 234.

Blind Watchmaker, but others may prefer to remove their dark glasses."[58] Mivart had clearly removed his dark glasses in the nineteenth century even as Darwin held to the Blind Watchmaker. And Conway Morris has removed his in the twenty-first. And he may not be alone. Even the great Darwinian apologist Ernst Mayr seems to have misplaced his dark glasses late in life. Earlier in his career, Mayr had argued vehemently against comparing natural selection to the work of an engineer, due to the obvious teleological implications. For Mayr a preferred figure of speech for the evolutionary process came from François Jacob: the evolutionary process works like a tinkerer. But in his nineties, when commenting on the phenomenon of convergent evolution, Mayr wrote, "Convergence illustrates beautifully how selection is able to make use of the intrinsic variability of organisms to *engineer* adapted types for almost any kind of environmental niche."[59] If evolution works like an engineer, evolution is purposeful. Mivart recognized the profound challenge evolutionary convergence represented for Darwin's theory. More than a century later, the riddle of evolutionary convergence remains.

All of Conway Morris's examples of convergence come from the animal world. So what about plants? Do they exhibit convergence too? George Ledyard Stebbins, the biologist who would be credited with bringing botany into the fold of the modern evolutionary synthesis during the 1940s, identified evolutionary trends that are "repeated so many times in various groups of plants that some consideration of them is essential to an understanding of the mechanism of evolution."[60] Given the existence of these repeated trends, Stebbins believed the conclusion was inevitable that "evolutionary trends must be governed by some guiding force."[61] For Stebbins, the force guiding evolution is either natural selection (a process dependent on surviving species adapting to environmental factors) or orthogenesis (a process not dependent on species' relations to environmental factors)—and Stebbins doesn't side with orthogenesis: "The conclusion is inevitable . . . that all long-continued evolutionary trends in plants, as well as in animals, are guided by natural selection."[62]

---

58. Conway Morris, *Life's Solution*, 330.
59. Mayr, *What Evolution Is*, 223 (italics added).
60. Stebbins, *Variation and Evolution in Plants*, 476.
61. Ibid., 500.
62. Ibid., 501.

But Stebbins undercuts his own certainty in natural selection as evolution's guiding force. He observes that some living plants exhibit primitive features (from early stages of evolution) whereas others show more advanced features (from later stages of evolution). If natural selection were evolution's guiding force, then plants that exhibit primitive features could not coexist with plants that show advanced features. Stebbins is thus forced to conclude: "The differentiation of orders and families of flowering plants through the action of natural selection under present conditions is well-nigh impossible to imagine."[63] How it is possible to consider a conclusion inevitable that is simultaneously "well-nigh impossible to imagine" is itself hard to imagine. Clearly, evolutionary convergence in plants presents as big a challenge to Darwinian evolution as does its evolutionary convergence in animals. Convergence is a mysterious feature of the entire organic world, and no less so is the repeated evolution of an organ of extreme complexity like the visual apparatus.

Convergence and the Evolution of the Eye

One of the most startling forms of convergence is represented by the evolution of the eye. Many biologists think the complex camera-like eyes found in vertebrates and cephalopods (squid, octopus, and the like) may have evolved independently over sixty times.[64] How could the same random mutations occur repeatedly so many times as to allow for the repeated evolution of such an intricately complex organ of sight? Even Darwin confessed to Asa Gray, "The eye to this day gives me a cold shudder."[65] And upon reading the page proofs of the *Origin* before publication, Charles Lyell gave Darwin the following editorial advice:

> The first page of this most important summary gives the adversary an advantage by putting forth so abruptly and crudely such a startling objection as the formation of "the eye" not by means analogous to human reason or rather to some power immeasurably superior to human reason but to superinduced variations like those of which a cattle breeder avails himself. Pages would be required thus to state an objection and remove it—it would

---

63. Ibid.

64. Salvini-Plawen and Mayr, "On the Evolution of Photoreceptors and Eyes," 247.

65. Letter from Darwin to Gray dated October, 1860. In Darwin, ed., *Life and Letters*, vol. 2, 273.

be better as you wish to persuade to say nothing? Leave out several sentences and in a future edition bring it out more fully.[66]

Lyell knew how controversial Darwin's discussion of eye evolution would be and thought he would be better off simply omitting it until he could make a more compelling argument in a future edition.

Mivart also found the eye perplexing. Because of its complexity, he could not fathom how it could evolve in a gradual Darwinian fashion when a small change in only one part at a time would almost certainly render it useless. Mivart argued that any higher organ of extreme complexity could evolve only through the process of multiple coordinated variations occurring simultaneously in all its interdependent parts, and the necessity of multiple coordinated variations went diametrically against the Darwinian view that variations arise randomly without reference to their potential usefulness in a particular environmental context. Mivart wrote:

> It is not, for instance, in the strictest sense of the word, impossible that a poem and a mathematical proposition should be obtained by the process of shaking letters out of a box; but it is improbable to a degree that cannot be distinguished from impossibility; and the improbability of obtaining an improvement in an organ by means of several spontaneous variations, all occurring together, is an improbability of the same kind.[67]

Mivart even tried to quantify his objection. If a single variation occurs one in a thousand times, and it takes ten coordinated variations to bring about an improvement that could be preserved by natural selection, the odds of this happening would be one in $1,000^{10}$, or one in 10^{30}, "a number about ten thousand times as great as the number of waves of light that have fallen on the earth since historical time began."[68] It seems highly improbable that randomly generated variation acted on by natural selection could lead to the evolution of the eye even once, let alone more than sixty times.

Despite his cold shudder, Darwin did try to meet this challenge in the last edition of the *Origin*. Arguing that a person who finishes reading the *Origin*, who acknowledges that a large body of formerly inexplicable

---

66. Letter from Lyell to Darwin dated Oct. 3, 1859. In Burkhardt et al., eds., *Correspondence*, 7:340.

67. Mivart, *On the Genesis of Species*, 65.

68. Ibid.

facts can now be explained by the theory of natural selection, "ought to admit that a structure even as perfect as an eagle's eye might thus be formed, although in this case he does not know the transitional states,"[69] Darwin recognizes an objection: to remain functional the eye would have to undergo multiple simultaneous variations at each stage of its evolution. But he responds, "As I have attempted to show in my work on the variation of domestic animals, it is not necessary to suppose that the modifications were all simultaneous, if they were extremely slight and gradual."[70] Darwin provides no evidence to support this and ignores the problem that if variations are too slight they might not provide the kind of advantage necessary to be preserved by natural selection in the first place. Recognizing the weakness of his response, Darwin concludes:

> To arrive, however, at a just conclusion regarding the formation of the eye, with all its marvelous yet not absolutely perfect characters, it is indispensable that the reason should conquer the imagination; but I have felt the difficulty far too keenly to be surprised at others hesitating to extend the principle of natural selection to so startling a length.[71]

As in so many other cases, Darwin ultimately concedes to his critics. He knew he lacked the evidence to convince them of something that seemed so obvious to him.

More recently, the outspoken Darwinian evolutionary biologist Jerry Coyne has picked up where Darwin left off, explaining to a general readership why the evolution of the eye does not pose a problem for Darwinian evolution. In his book *Why Evolution Is True*, a book designed to put criticisms of Darwinian theory to rest, Coyne argues that the development of the eye's complexity "can be broken down into a series of small, adaptive steps."[72] His claim rests on the widely cited work of Swedish researchers Dan-Eric Nilsson and Susanne Pelger, who developed a model showing how a light-sensitive patch of cells backed by a pigment layer could develop into a complex camera eye in a relatively short period of time. Nilsson and Pelger's model calculates that this could happen through a series of 1,829 tiny adaptive steps taking no more than four hundred thousand years, a geologically brief span of time.

69. Darwin, *On the Origin of Species*, 134.
70. Ibid.
71. Ibid., 135.
72. Coyne, *Why Evolution Is True*, 142.

Since eyes are attested in the fossil record back to 550 million years ago, ample opportunity existed for the repeated evolution of complex eyes. In Coyne's view, Nilsson and Pelger's work closes the deal on eye evolution. But in his rush to convince readers of the strength of Darwinian theory, Coyne has ignored important caveats in Nilsson and Pelger's work.

In the seminal article to which Coyne is referring, Nilsson and Pelger, after laying out their model in detail, ask a very pointed question. If complex eyes can evolve as fast as their model predicts, why do we find so many examples of intermediate eye designs among currently living animals? Shouldn't they all have evolved to full, complex form? The answer to this riddle lies, according to Nilsson and Pelger, in

> a fact that we have *deliberately ignored*, namely that an eye makes little sense on its own. Although reasonably well-developed eyes are found even in jellyfish, one would expect most lens eyes to be useless to their bearers without advanced neural processing. For a sluggish worm to take full advantage of a pair of fish eyes, it would need a brain with large optic lobes. But that would not be enough, because the information from the optic lobes would need to be integrated in associative centres, fed to motor centres, and then relayed to the muscles of an advanced locomotory system. In other words, the worm would need to become a fish. Additionally, the eyes and all other advanced features of an animal like a fish become useful only after the whole ecological environment has evolved to a level where fast visually guided locomotion is beneficial.[73]

Mivart could not have said it better. Each stage of eye evolution would likely have been useless to an organism absent a complex array of simultaneous variations occurring in the brain, allowing the organism to make use of the light being collected and focused by the developing eye.

We do not really see with our eyes. Eyes merely collect and focus light. Vision is a subjective experience created by our brains as they interpret the signals coming from the eye via the optic nerve. One can have fully functioning eyes and still be blind if the visual centers of the brain are compromised by illness or injury. Given this, Nilsson and Pelger conclude, "Because eyes cannot evolve on their own, our calculations do not say how long it *actually* took for eyes to evolve in the various animal groups. However, the estimate demonstrates that eye evolution would be extremely fast if selection for eye geometry and optical structures

---

73. Nilsson and Pelger, "Pessimistic Estimate," 58 (italics added).

imposed the only limit."⁷⁴ But Nilsson and Pelger have said that geometry and optical structure do not impose the only limit. For this reason they assume that primitive eyes continue to exist among living creatures because the conditions of their life do not require more advanced visual processing. A worm does not need fish eyes, simply because it is not a fish. But by pointing out the need for complex, coordinated variations to make an eye evolve to whatever level of complexity is needed by a particular organism, Nilsson and Pelger's model cannot necessarily be used to prove that eyes could evolve in a Darwinian way. Like Darwin, they show how in theory it *might* happen but certainly do not prove how it *did* happen. Darwin's cold shudder had a firm basis. More important, however, Coyne's selective citation of Nilsson and Pelger's work demonstrates how the grand narrative of Darwinian triumph obscures the ambiguous nature of Darwinian theory when it is packaged and sold to the larger public, a dynamic we will meet many more times moving forward.

By 2012, Pavel Vopalensky and his collaborators could still say, "The evolutionary origin of vertebrate eyes is enigmatic."⁷⁵ This is due to the paucity of existing organisms showing the useful gradations in improvement of the eye that must have occurred during evolutionary history. Since complex eyes are documented in the fossil record back as far as 530 million years ago during the Cambrian period, earlier stages of eye evolution must have occurred in simpler pre-Cambrian organisms. Clues to the evolution of the complex vertebrate eye then may lie buried in the physiology of today's simpler organisms. This reasoning prompted Vopalensky and his colleagues to study the more primitive eye of amphioxus, a small fish-like chordate (organism with a spinal cord) that lives in the world's oceans. They show that the primitive eye of amphioxus is biochemically similar to aspects of the vertebrate eye and conclude: "The amphioxus frontal eye circuit thus represents a very simple precursor circuit that, by expansion, duplication, and divergence, might have given rise to photosensory-locomotor circuits as found in the extant vertebrate brain."⁷⁶ They provide no evidence for how this "expansion, duplication, and divergence" took place and whether the process can be explained based on Darwinian principles, but it is this "how" question that is what makes the evolution of the vertebrate eye such an enigma. Thus,

---

74. Ibid. (Italics added).
75. Vopalensky et al., "Molecular Analysis of the Amphioxus Frontal Eye," 15383.
76. Ibid., 15387.

the title of their paper—"Molecular Analysis of the Amphioxus Frontal Eye Unravels the Evolutionary Origin of the Retina and Pigment Cells of the Vertebrate Eye"—seems a bit disingenuous. If they cannot show the mechanism by which this evolutionary trajectory took place, they can hardly claim to have "unraveled" the mystery.

In 2013, Dan-Eric Nilsson reentered the eye evolution debate with an article showing how different steps in eye evolution are coordinated with new behavioral patterns that might increase fitness and thus be subject to natural selection. He begins by stating the problem that so vexed Mivart: "One of the first steps in eye evolution must have been the appearance of a light dependent chemical reaction coupled to a signaling system."[77] Obviously, as previously said, a light-sensitive patch would prove useless to an organism that did not simultaneously develop a neural processing system that could create the subjective experience of sight. It is the coordination of these two (light sensitivity and neural processing together) that makes the evolution of the eye such an enigma, at least for a Darwinian process. Nilsson then details four possible different levels of behavior that become possible when advances in visual processing move from nondirectional photoreception (the simple ability to distinguish light from darkness) to the high-resolution vision enjoyed by advanced vertebrates. In his conclusion, he observes that "For each evolutionary transition to a higher class, the rate of information delivered to the nervous system increases by several orders of magnitude. This provides opportunities for more advanced behaviours and requires larger brains."[78] But he says nothing about how this larger brain could evolve in a coordinated way with the evolution of the eye so as to be able to process effectively this enormous increase in information at each step of the evolutionary process, from nondirectional photoreception to high-resolution vision. This is the key question at the heart of eye evolution, and, Nilsson's efforts notwithstanding, why eye evolution remains very much a mystery.

Does this mean eye evolution, and evolution in general, must be directed by some supernatural force? Some would say yes, but other naturalistic explanations are available. Robert G. B. Reid was an outspoken critic of Darwinian evolution but no ally of creationists or ID supporters. Reid argued that eye evolution might best be explained by recourse to

---

77. Nilsson, "Eye Evolution and Its Functional Basis," 5.
78. Ibid., 18.

epigenetic inheritance systems. Epigenetic inheritance describes the way characteristics can sometimes vary and be passed on without involving DNA base pair substitutions (or mutations). We will learn more about epigenetic inheritance in a later chapter. For now, what is important is to observe how Reid draws on epigenetic research to argue that eyes can evolve via "ubiquitous generative conditions. Their emergent transition into a more organized and more energetically stable, phase requires no directing selection pressure."[79] Determining whether or not Reid is correct is less important than observing how a member of the biological establishment—Reid was Professor of Biology at the University of Victoria in British Columbia—felt that the evidence for a Darwinian explanation for eye evolution was ambiguous enough for him to propose a scientific alternative. If Darwinian evolution were as well supported empirically as the grand narrative portrays, no biologist of Reid's stature would feel the need to propose an alternative theory or would spend valuable research time and money on such a fruitless enterprise. But members of the biological establishment *do* offer alternative scientific theories to explain evolutionary phenomena more frequently than most people realize, as we will see in chapter 7.

Given the ambiguous nature of the empirical support for Darwinian evolution, its defenders often feel compelled to engage in rhetorical rather than scientific arguments when debating their creationist opponents. If we assume that God or some other intelligent force designed the eye as the perfect organ of sight, we run into the problem that the eye does not appear to be as perfect as we think. The vertebrate eye, including in humans, contains an inverted retina. Rather than pointing toward the light coming through the lens, photoreceptors in the retina point away from the light. As a result, the optic nerve coming off the back of the retina has to loop around and pass through a hole in the retina to make its way to the brain. This results in all vertebrate eyes having a blind spot in the center of the field of vision. Darwinian defenders are fond of pointing out how ham-handed this arrangement is. No intelligent creator designing the eye from scratch would do it this way. As Richard Dawkins says, an engineer "would laugh at any suggestion that the photocells might point away from the light, with their wires departing on the side *nearest* the light."[80]

79. Reid, *Biological Emergences*, 249.
80. Dawkins, *Blind Watchmaker*, 93.

Of course it is not a scientific argument to pretend to know what an intelligent designer would or would not do. But a scientific argument against the idea of poor eye design has recently been made. In a 2009 edition of the journal *Vision Research*, Ronald H. H. Kröger and Oliver Biehlmaier make a case for the benefits of an inverted retina, especially in smaller animals. They call it a space-saving solution that likely facilitated the evolution of complex vertebrate eyes:

> Instead of asking why vertebrates possess apparently problematic inverted retinas, one may ask why such space-saving retinas are limited to vertebrates and a handful of invertebrates. The answer is the same for both questions: animals have their group-specific eye and retina types because of common descent within each phylogenetic group. Vertebrates have evolved into the group of animals which most heavily rely on vision with high spatial resolution. The inverted retina has most likely been an important factor since it allows for massive retinal processing of visual information without investment of precious space and weight.[81]

Perhaps Dawkins's engineer is no longer laughing. What appears to be a design flaw to one pair of eyes looks more like an economical solution to another. This of course does not prove that the eye is intelligently designed, but it does illuminate the flawed and tendentious nature of the zeal with which antidesign crusaders attempt—rhetorically, rather than scientifically—to undermine design arguments by appealing to hypothetical scenarios about how a designer would or would not act.

Despite the real difficulties one encounters trying to account for the evolution of the eye based on Darwinian principles, we continue to get statements like this one from H. Allen Orr:

> The only thing that 'guides' evolution is sheer, cold demographics. If a worm with a patch of light-sensitive tissue leaves a few more kids than a worm that cannot tell if the lights are on, that is where evolution will go. And, later, if a worm with light-sensitive tissue and a rough lens escapes a few more predators, that is where evolution will go.[82]

A more circumspect approach to eye evolution seems to be warranted based on the foregoing discussion. We would do well to recall the

---

81. Kröger and Biehlmaier, "Space-Saving Advantage," 2320.
82. Orr, "Response to Berlinski," 286.

amazement registered by the Spanish Nobel laureate Santiago Ramón y Cajal, widely considered to be the father of modern neuroscience:

> As the reader will remember, my devotion to the retina is ancient history. The subject always fascinated me because, to my idea, life never succeeded in constructing a machine so subtilely devised and so perfectly adapted to an end as the visual apparatus... I must not conceal the fact that in the study of this membrane I for the first time felt my faith in Darwinism (hypothesis of natural selection) weakened, being amazed and confounded by the supreme constructive ingenuity revealed not only in the retina and in the dioptric apparatus of the vertebrates but even in the meanest insect eye. There, in fine, I felt more profoundly than in any other subject of study the shuddering sensation of the unfathomable mystery of life.[83]

It is for good reason that Darwin confessed to Asa Gray that the eye gave him a cold shudder. Darwin might still be shaking today.

Charles Darwin died on April 19, 1882, having garnered enough fame to warrant a burial in Westminster Abby. Publication of *On the Origin of Species* had indeed catapulted him to a place of prominence among nineteenth-century scientists. But it really cannot be said that he formulated the greatest idea anyone has ever had. He did not spark a revolution in thought on the nature of biological evolution. At the end of the day, Charles Lyell and Asa Gray continued to hold to teleological understandings of evolution. Thomas Huxley, while working tirelessly to secure a wide reading for the *Origin*, made it clear that he still questioned the truth value of natural selection. And Alfred Russel Wallace, the independent codeveloper of the theory of natural selection, abhorred the notion that natural selection could explain the evolution of the human mind, and he gave himself over to a life of séances and spiritualism. For all the fanfare, the *Origin* largely failed to spark a revolution in evolutionary thought, and most scientists in the late nineteenth century continued to hold to Lamarckian, saltational, or vitalistic theories of evolutionary mechanisms.

Darwin is no doubt an important figure in the history of science. And we should not lose sight of his strengths as a scientist: his great powers of observation and his patience in exhaustively studying the most seemingly insignificant organisms—barnacles, pigeons, and

---

83. Ramón y Cajal, *Recollections of My Life*, 576n4.

earthworms—searching for the secrets they held for understanding the history of life on earth. But Darwin was also a deeply flawed scientist, capable of making intuitive leaps of imagination based on the flimsiest evidence. His attempts to make great contributions to geology largely misfired. And the hesitancy of his contemporaries to read the *Origin* as a compelling argument for natural selection was based simply on the lack of evidence Darwin had marshalled for his theory. As the nineteenth century gave way to the twentieth, Darwin's theory languished. Yet before long it would come to be advanced by many as *the* theory of evolution, indeed as the *only* theory of evolution. How did this stunning and unexpected transformation occur? The next few chapters will seek to provide an answer.

3

The Eclipse and Recovery of Darwinism

IN THE YEARS IMMEDIATELY following Charles Darwin's death in 1882, the evolutionary theory ascribed to him had largely fallen into eclipse. *On the Origin of Species* continued to be read and studied by the educated classes in both England and elsewhere, but most readers found the argument for natural selection generally unconvincing. Historian Peter Bowler reminds us of the importance of realizing how "anti-Darwinian theories were once taken seriously by large numbers of professional biologists. These scientists may have been influenced by moral or religious beliefs, but they were able to provide empirical arguments to support their position."[1] The grand narrative of Darwinian triumph tends to ignore or explain away this historical period in order to obscure the pervasive nature of the anti-Darwinian sentiments expressed by very smart and highly educated intellectuals in the late nineteenth and early twentieth centuries. Today, criticism of Darwin often marks one as unintelligent and uneducated, but the roster of turn-of-the-century Darwinian critics reads like a who's who of late Victorian intelligentsia.

The famous Irish playwright George Bernard Shaw was an outspoken critic of natural selection and held to a form of what he called "creative evolution."[2] Darwin's own cousin, Francis Galton, father of the early eugenics movement, held to a saltational view of evolution. Even John Henslow, the Cambridge botanist who had recommended Darwin for the *Beagle* voyage, could not understand how the struggle for existence

---

1. Bowler, *Eclipse of Darwinism*, 10.
2. See especially his play *Back to Methuselah*.

would not eliminate most of its victims at an early age, before adaptive characters were formed. And this rejection of Darwinism was not happening only in Darwin's home country. Anarchist philosopher Peter Kropotkin, in Russia, believed that organisms could respond in a purposeful way to their environmental contexts, rendering selection unnecessary. And Henri Bergson, in France, coined the term *élan vital* to describe the innate force he believed drove evolution in particular directions. In fact, Darwinism never really did gain a foothold in France: by one estimate 95 percent of French biologists and philosophers still rejected Darwinian evolution as late as the 1970s.[3] One noteworthy exception to this anti-Darwinian trend was Ernst Haeckel, in Germany. Haeckel was an aggressive supporter of natural selection, but his motivations may have been more political than scientific. Haeckel was a strong supporter of German nationalism and saw in natural selection a scientific basis for racial purification. As Simon Conway Morris has said about Haeckel, "Behind the bearded sage and devotee of the little town of Jena was an intolerant mind wedded to racism and antisemitism."[4]

Despite Haeckel's prejudices, he is still an important figure in the history of evolutionary theory, for he created a series of drawings comparing a variety of vertebrate embryos at three stages of their development, including what he called their earliest stage. Noting that the embryos all look very similar at the earliest stage and then diverge into their separate forms as they develop, Haeckel believed he had found evidence for evolution—vertebrate embryos look very similar early and then diverge as they develop because they all went through a similar evolutionary history as they evolved from a common ancestor. Haeckel dubbed this theory "ontogeny recapitulates phylogeny." That is, the evolutionary history of an organism (its phylogeny) is recapitulated during its embryonic development (its ontogeny). Even human embryos, Haeckel observed, seem to develop gill slits that later disappear during embryonic development, demonstrating their evolution from marine organisms. Haeckel's embryo drawings quickly became famous and have been passed down to modern times as evidence for evolution.[5] They have become what Jonathan Wells calls an "icon of evolution."

---

3. Boesiger, "Evolutionary Biology in France," 309.
4. Conway Morris, *Life's Solution*, 319.
5. For the story of Haeckel's embryos, see Hopwood, *Haeckel's Embryos*.

Early on, however, Haeckel's embryo drawings were shown to be inaccurate, and he was charged with forgery. Whether Haeckel intentionally misrepresented the embryos is unclear. What is clear is that he drew them to look more alike than they actually do, and what he called the earliest stage of development is really an intermediate stage; the embryos look quite different at their true earliest stage of development. Indeed the parallel lines that show up on the necks of all vertebrate embryos at this intermediate stage—what Haeckel identified as gill slits—are now identified as pharyngeal arches. They turn into gills in fish embryos but develop into quite other structures in other vertebrates. They are not direct evidence that all vertebrates evolved from fish. Clearly, Haeckel's desire to champion natural selection in service to his racial ideology led him to misrepresent (intentionally or unintentionally) potential evidence for the theory. But the story of Haeckel's embryos does not end there.

The grand narrative of Darwinian triumph co-opted these iconic images and has continued to perpetuate them as evidence for evolution despite their widely documented inaccuracies. Ernst Mayr's 2001 book *What Evolution Is* includes a picture of Haeckel's embryo drawings with the caption, "Embryos of different vertebrates at three comparable stages of development. The earlier the stage of development the more similar are the embryos of different groups."[6] Mayr does not refer to Haeckel but instead cites a 1990 textbook as the source of the drawings. We are told nothing about the inaccuracies in the drawings or about the fraud charges leveled at Haeckel. Indeed, we are told nothing about Haeckel at all! Similarly, Gregory A. Wray, in an article in *Evolution: The First Four Billion Years*, published by Harvard University Press in 2009, reproduces Haeckel's embryo drawings when discussing evolution and development. Wray at least credits the drawings to Haeckel, but his caption reads:

> His famous illustrations of vertebrate embryos, one of which is shown here, depict progressive divergence in overall similarities among species as development proceeds . . . Note that all the embryos have gill slits at an early stage and that the human embryo has a tail that is later lost. It was the presence of these phylogenetically ancient features in embryos, despite their absence in adults, that strongly impressed Haeckel.[7]

---

6. Mayr, *What Evolution Is*, 28.
7. Wray, "Evolution and Development," 211.

Again, there is no discussion of the well-known inaccuracies in Haeckel's drawings, and the discredited gill-slit theory is perpetuated. These are just two examples of many that could be cited. Haeckel's century-old embryo drawings were a staple in biology textbooks for generations and still appear in some textbooks even today, even though high-resolution digital images that accurately portray the comparison between embryos are widely available. Just as Haeckel's ideological interests may have led him to misrepresent this piece of "evidence" for evolution, so the shapers of the grand narrative have unfortunately continued the misrepresentation in service to their own ideological commitments to Darwinism.[8]

Haeckel's ideologically motivated support of natural selection aside, Darwinian theory remained a minority opinion around the turn of the century. But eclipses are transitory phenomena, and by the 1920s the eclipse of Darwinism would begin to wane. Darwin's theory of natural selection would find itself on the way to becoming synonymous with the very concept of evolution. How did this transition occur? We might assume that compelling new evidence came to light forcing the emerging field of biology to recognize that Darwin had been right all along. But we must heed Peter Bowler's warning:

> It no longer seems quite so obvious that a successful new theory triumphs merely because of its greater factual support . . . The process by which a new theory gains its dominant status is a complex one, involving not only experimental testing but also shifting philosophical concerns and changes in the structure of the scientific community.[9]

For his part, Bowler clearly believes that the reemergence of Darwinism constituted the reemergence of a true theory that had simply lain dormant for a time. But I will argue that the reemergence of Darwinism is much better explained by "shifting philosophical concerns and changes in the structure of the scientific community" than it is by the discovery of new empirical evidence.

---

8. For additional examples of distorted evidence for evolution in textbooks, see Wells, *Icons of Evolution*.

9. Bowler, *Eclipse of Darwinism*, 11.

The Rediscovery of Mendelian Heredity

Darwin was working with a considerable handicap when developing his theory of natural selection. Based on his painstaking observations, he recognized in a way few others had that the individual members of a population of organisms were not exact replicas of one another. They varied in innumerable small details. What he did not know was how these variations were produced or how they were inherited. His critics seized on this lacuna in his knowledge with abandon. If an individual was born that just so happened to possess a variation rendering it more fit in its environment, it would still be forced to mate with an individual lacking this variation. Most people in Darwin's day believed in some form of blending inheritance. The progeny of this mating would therefore be born with only about half the variation of the more fit parent. In future generations, this variation would be blended away even further until it was lost altogether. As long as parental characteristics get blended together, there is no way to conceive how natural selection could possibly lead to the evolution of new characteristics and whole new organisms. Unfortunately for Darwin, a potential answer to this dilemma had been discovered during his lifetime, but he was unaware of it.

In the 1860s, the Moravian monk Gregor Mendel experimented with pea plants in the monastery garden. He crossed plants that produced different varieties of peas, keeping track of the distribution of pea characteristics in succeeding generations. Mendel discovered that blending inheritance was wrong. Full parental variations appeared in future generations in predictable proportions, leading him to propose the theory of particulate inheritance. Rather than being blended away, parental traits are inherited in discrete units (later called genes by Wilhelm Johannsen): an organism possessing a variation rendering it more fit in its environment could mate with a less fit organism but still pass its more fit characteristic to some of its offspring. Unfortunately for Darwin, Mendel's groundbreaking research was published in an obscure journal and made little impact in its day. Darwin died never knowing about Mendel's potential correction to his theory's near-fatal flaw.

This situation changed, however, right at the turn of the century. Dutch botanist Hugo de Vries had been experimenting with the evening primrose, crossing varieties in much the same way Mendel had done with his pea plants. De Vries's experiments produced similar results, and in 1900, de Vries rediscovered Mendel's obscure paper and brought the theory of particulate inheritance into the scientific mainstream. With this new quantification of the laws of heredity, we might expect that biologists would have rushed to apply this knowledge to Darwin's theory. But this was not to be. As de Vries continued to hybridize his evening primrose plants, he discovered that whole new varieties unexpectedly emerged with characteristics that could not be traced back to parental stocks. He dubbed this sudden emergence of new varieties "mutation" and published a new explanation of evolution he called *mutationstheorie*. De Vries believed evolution proceeded in jumps, or saltations, driven by these large mutational changes and leading to the sudden emergence of whole new varieties of organisms. The rediscovery of Mendelian genetics, rather than being put to the service of Darwinism, became just one more reason to question Darwin's gradual conception of the evolutionary process.

De Vries was not the only one who viewed Mendelian heredity as contrary to Darwin's gradualist understanding of evolution. The early British geneticist William Bateson agreed. Even before he became aware of Mendel's work, Bateson had carried out a series of experiments during an eighteen-month sojourn in Central Asia during the years 1886–1887. To test both Lamarck's and Darwin's views of evolution, he collected and studied organisms from in a series of lakes varying in characteristics like size, depth, and salinity. Bateson wanted to see if variations in organisms could be correlated to variations in the environment. Lamarck's theory taught that the environment drove evolutionary change directly, whereas Darwin's taught that natural selection had the effect of adapting organisms to their environment. In either case, Bateson believed, variations in organisms should track variations in the environment since the environment was understood to influence evolution, whether that influence was direct (Lamarck) or indirect (Darwin). Bateson found, however, that no discernable correlation existed between organismal and environmental variation. Organisms seemed to vary in ways that apparently had no connection to changes in their environment. This undermined, in Bateson's view, both Lamarckism and Darwinism and led Bateson to focus solely on the role of variation in evolution. He became convinced that "selection did not play a role in determining whether a new form could become established; it was variation itself that was central in directing the course of evolution."[10]

---

10. Bowler, *Eclipse of Darwinism*, 190.

Bateson further observed that environmental contexts tend to change in slow, gradual ways, but species often vary from one another in discontinuous ways. That is, individual organisms can easily be placed in one species category or another. We don't find organisms displaying a gradual continuum from one species to another. All squirrels are clearly squirrels. All chipmunks are clearly chipmunks. But we don't find rodents running about in our backyard that appear ambiguous as to whether they are squirrels or chipmunks. Species vary discontinuously while environments vary continuously. Bateson thus concluded that evolution was driven entirely by discontinuous variation, without any reference to environmental influence. In addition, Bateson observed that the characteristics that taxonomists use to distinguish one species from another frequently fail to show signs of adaptive utility—further evidence that natural selection played no role in the evolution of new species. Bateson published his anti-Darwinian (and anti-Lamarckian) views in a famous work, *Materials for the Study of Variation*, in 1894, in which he concluded, "if the problem of species is to be solved at all it must be by the Study of Variation."[11] When de Vries rediscovered and publicized Mendel's laws of heredity after 1900, Bateson was already well primed to agree with de Vries's saltational view of evolution. Evolution seemed to be driven by mutations, understood as sudden, large discontinuous changes spontaneously arising in the offspring of organisms.

In addition, Bateson echoed Mivart's earlier concern that multiple coordinated variations would be necessary to keep organisms functional during the evolutionary process, a clear problem for Darwinism if variations are produced randomly. In an 1888 letter to his sister Anna, Bateson wrote:

> My brain boils with evolution. It is becoming a perfect nightmare to me. I believe now that it is an axiomatic truth that no variation, however small, can occur in any part without other variation occurring in correlation to it in all other parts; or, rather, that no system, in which a variation of one part had occurred without such correlated variation in all other parts, could continue to be a system. This follows from what one knows of the nature of an "individual," whatever that may be.[12]

---

11. Bateson, *Materials for the Study of Variation*, 575.
12. Bateson, *William Bateson, FRS.*, 39.

Throughout a life steeped in the study of variation and heredity, Bateson saw no empirical reason to accept Darwinian evolution and many good reasons to reject it. Even as late as 1922 in an address given to the American Association for the Advancement of Science, Bateson would say: "In dim outline evolution is evident enough. From the facts it is a conclusion which inevitably follows. But that particular and essential bit of the theory of evolution which is concerned with the origin and nature of *species* remains utterly mysterious."[13]

Although de Vries and Bateson were influential figures, perhaps no one was more influential in the early twentieth century than the Nobel Prize-winning geneticist Thomas Hunt Morgan. Morgan won the Nobel Prize for his work on mapping the units of hereditary variation (genes) onto chromosomes. He was a native of Kentucky, but developed a strong aversion to the conservative Christianity of his upbringing, leading to efforts to remove any vestiges of religious thought from biology and to turn it into a truly scientific endeavor. At the same time, Morgan was critical of the proponents of Darwinism for developing speculative scenarios to explain how natural selection might work in a particular circumstance. Not only did Morgan think biology needed to become less religious, he also wanted to see it become more empirical and experimental.

It says much, then, about the state of Darwinian evolution in the early twentieth century that an antireligious and hardheaded empiricist like Morgan was still so critical of Darwinism. In his first teaching post, at Bryn Mawr College in the 1890s, Morgan even regularly taught a course on criticisms of Darwinism.[14] Morgan early accepted Bateson's idea of discontinuous variation and de Vries's mutation theory, and it was for the purpose of trying to find evidence for de Vriesian mutations that he began his famous work of breeding *Drosophila* (fruit flies) after moving to Columbia University. When he looked at Darwinism, Morgan failed to see how natural selection could produce anything new. He reasoned that breeding for height in a population might produce more tall people, but selection could not create a person any taller than the tallest in the original population. For Morgan, selection "has not produced anything new, but only more of certain kinds of individuals. Evolution, however, means producing new things, not more of what already exists."[15] If new

---

13. Ibid., 393.
14. Allen, *Thomas Hunt Morgan*, 53.
15. Morgan, *Evolution and Genetics*, 127.

characteristics were going to be introduced into a population, it could only be via a major mutational change.

Morgan's criticism of Darwinian evolution presents a real problem for the grand narrative of Darwinian triumph. From all appearances, Morgan should have been a poster child for Darwinism. He had access to the crucial piece of evidence that Darwin lacked—Mendel's theory of particulate inheritance. He was rigorously materialistic in his philosophical outlook and worked tirelessly to place the emerging field of biology on a firm experimental foundation. And as one of the most important scientists of the twentieth century, Morgan cannot be dismissed as a marginal figure. For his part, Ernst Mayr explains Morgan's anti-Darwinian position as a function of Morgan's supposed essentialism. Morgan, according to Mayr, was not a population thinker, so it was impossible for him to understand Darwin's theory.[16] This criticism rings hollow, however, as Morgan manifestly did engage in population thinking when he argued that selection could produce more taller people in a population but could not produce a person taller than the tallest one in the original population. Morgan understood how population theory worked. He just did not see how it could create anything new. Accusing Darwin's critics, even eminent scientists, of not understanding his theory is a common ploy. These early Mendelian geneticists simply did not see evidence to suggest that Darwin's slow, gradual evolutionary process could account for large-scale evolutionary change.

Even though a saltational view of evolution dominated the early twentieth century, Darwin's theory did have its defenders. In contrast to de Vries, Bateson, and Morgan, the German biologist August Weismann became a leading voice for natural selection around the turn of the century. In fact, Weismann was more Darwinian than Darwin himself, spawning a movement dubbed neo-Darwinism in order to distinguish between the two. Weismann became a strong critic of the Lamarckian idea of inheritance of acquired characters. He famously removed the tails of sixty-eight mice and then bred them over five generations. Eventually, 901 mice were produced but not one lacked a tail or even possessed an abnormally shorter tail. Weismann concluded that acquired characters could not be inherited, rendering Lamarckism moot. He theorized that early in development, the germ cells of an organism (egg and sperm) split off from the somatic cells (cells that make up the rest of the body

---

16. Mayr, "Prologue," 29.

tissues). If inheritance is transmitted through the germ cells, Weismann saw no way that changes in the somatic cells acquired during the life of an organism (acquired characters) could be passed down to future generations. Having disproven Lamarckism, Weismann then advanced natural selection as the sole driver of the evolutionary process. Since Darwin had retained a role for Lamarckism in his theory, we can see why Weismann was considered to be more Darwinian than Darwin himself. But on what grounds did Weismann base his fierce advocacy of natural selection?

In 1909, Weismann penned an essay titled "The Selection Theory" as a contribution to a volume of essays marking the fiftieth anniversary of the publication of the *On the Origin of Species*. In it he tackles one of the major questions surrounding the efficacy of natural selection: the selection value of the initial stages of variation. Darwin's theory requires that even the smallest, most incremental variations in an organism can have selection value—that is, will lead that organism to produce more offspring than others in the population. A tiny variation that lacks selection value will quickly disappear from the population, ending any emerging evolutionary trend. Weismann illustrates this with the example of the thickness of limpet shells that protect limpets from the force of the ocean waves breaking upon them. He asks, "What proportion of thickness was sufficient to decide that of two variants of a limpet one should survive, the other be eliminated?"[17] Weismann confesses to not knowing the answer to this question, but he assumes that some measure of thickness must have had selection value since limpets exist. So, can incremental variations really do the work that Darwin's theory requires of them?

Weismann writes:

> To this question even one, who like myself, has been for many years a convinced adherent of the theory of selection can only reply: *we must assume so, but we cannot prove it in any case*. It is not upon demonstrative evidence that we rely when we champion the doctrine of selection as a scientific truth; we base our argument on quite other grounds.[18]

We need not speculate about the "quite other grounds" on which Weismann bases his championing of natural selection. He explicitly names them later in the essay:

---

17. Weismann, "Selection Theory," 25.
18. Ibid. (Italics original).

> That selection is a factor, and a very powerful factor in the evolution of organisms, can no longer be doubted. Even although we cannot bring forward formal proofs of it *in detail*, cannot calculate definitely the size of the variations which present themselves, and their selection-value, cannot, in short, reduce the whole process to a mathematical formula, yet we must assume selection, because it is the only possible explanation applicable to whole classes of phenomena, and because, on the other hand, it is made up of factors which we know can be proved actually to exist, and which, *if* they exist, must of logical necessity cooperate in the manner required by the theory. *We must accept it because the phenomena of evolution and adaptation must have a natural basis, and because it is the only possible explanation of them*).[19]

Natural selection is for Weismann a default assumption that seems, to him at least, the only explanation for an evolutionary mechanism that can be squared with a purely naturalistic philosophy. (Morgan would have disagreed.) Weismann's argument here is an early form of the argument encountered in chapter 1 (above) where Douglas Futuyma and Stephen Stearns, writing in the recently published *Princeton Guide to Evolution*, call natural selection's greatest strength its ability to render biology a science like chemistry and physics. Weismann had no more evidence for natural selection than did the Medelians who rejected the theory. By his own admission he mainly accepts it for philosophical reasons.

This did not sit well with the empirically minded Morgan. He had little respect for the neo-Darwinian movement spawned by Weismann:

> They [the neo-Darwinians] seem less concerned with the advancement of the study of evolution than with expounding Darwinism as dogma. Their credulity is more remarkable than their judgment. To *imagine* a use for an organ is for them equivalent to *explaining* its origin by natural selection without further inquiry.[20]

It would be hard to disagree with Morgan given Weismann's surprisingly frank admission that his support for natural selection was not based on evidence but on "quite other grounds." The great Darwinian champion at the turn of the twentieth century was perpetuating a philosophical dogma, not a firmly supported scientific theory. The more empirically

---

19. Ibid., 61 (italics original).
20. Morgan, "For Darwin," 373 (italics original).

minded scientists of the era such as de Vries, Bateson, and Morgan could not square natural selection with the evidence at hand.

Despite the anti-Darwinism of the Mendelians, Darwin's theory would eventually emerge from its eclipse over the next couple decades. How exactly did this occur?

Population Genetics and the Recovery of Darwinism

For the Mendelians, the discontinuous character of particulate inheritance led naturally to the idea that evolution must proceed in a discontinuous manner as well. But other geneticists began experimental programs seeming to call this basic assumption into question. Chief among these was William Castle, at Harvard, who ran selection experiments on coat color in guinea pigs and rats. Castle had originally been a Mendelian heavily influenced by the work of William Bateson, but his experiments with coat color caused him to question assumptions made by the Mendelians about discontinuous evolution. Coat color gradations did in fact seem to form a continuous spectrum, and further work began to undermine the earlier assumption of a one-to-one correspondence between a gene and a particular phenotypic character. Geneticists discovered the twin concepts of *pleiotropy* (that a single gene can affect more than one trait) and *epistasis* (that a single trait can be influenced by more than one gene). It became clear that genes interacted with one another in complex ways, leading to the possibility that natural selection could change a character beyond its original limit of variation and in a smooth continuous way. Geneticists began to think more in terms of the genetics of populations of organisms rather than in terms of the genetics of individual organisms—an idea that gave birth to the discipline of population genetics.

Population genetics as a field of inquiry was essentially developed by three mathematically gifted individuals during the 1920s: Ronald A. Fisher and J. B. S. Haldane in Great Britain, and Sewall Wright in the United States. The grand narrative of Darwinian triumph often presents this trio as contributing to the development of a single theory capable of demonstrating the power of natural selection to account for evolution. But the reality is far more complex. I will briefly consider the work of each individual.

Ronald Fisher

Ronald Fisher (1890–1962) was a mathematician by training who turned his powerful quantitative skills to the problem of modeling how gene frequencies change over time in a population of organisms that is subject to selection pressure. He criticized Bateson for failing to recognize the mathematical aspects of biology and the way Mendel had supplied the missing parts of Darwin's theory. Fisher became convinced that natural selection was the sole driving force of evolutionary change, a fact he believed he had demonstrated through his sophisticated population genetic models. But questions arise about Fisher's true motivation for championing natural selection so forcefully.

One trait Fisher shared with many other geneticists of the early twentieth century (including Darwin's cousin Francis Galton) was an abiding interest in the emerging field of eugenics. If selection drives evolution, then it should be possible through the judicious use of artificial selection to improve the human race. Despite the obvious ethical problems inherent in selective breeding of humans, many early geneticists were proponents of eugenics programs. Peter Bowler humorously reports that Fisher's wife left him because he wanted to sire many children in order to perpetuate his own superior genes![21] Michael Ruse even suggests that Fisher created the field of population genetics for the express purpose of trying to underpin his hereditarian thoughts about society.[22] Although we should not ignore Fisher's interest in eugenics and its influence on his work, another motivation comes through clearly in his writings.

Fisher published his seminal text *The Genetical Theory of Natural Selection* in 1930. In the preface, Fisher argues that the chief attraction of natural selection to Darwin and Wallace was its ability to explain evolution based on "'known,' or independently demonstrable, causes." According to Fisher, alternative theories all rely on "hypothetical properties of living matter which are inferred from the facts of evolution themselves." We ought rather to "consider the theory of natural selection on its own merits."[23] In the body of the work, Fisher challenges the Mendelians with the observation that "For mutations to dominate the trend of evolution it is thus necessary to postulate mutation rates immensely greater than those which are known to occur and of an order of magnitude which

---

21. Bowler, *Evolution*, 311.
22. Ruse, "Dobzhansky and the Problem of Progress," 238.
23. Fisher, *Genetical Theory of Natural Selection*, vii.

in general would be incompatible with particulate inheritance."[24] All theories, Fisher argues, that rely on hypothetical mechanisms to control the occurrence of mutations or that direct the course of evolution must be set aside once the blending theory of inheritance is abandoned. He continues: "The sole surviving theory is that of natural selection and it would appear impossible to avoid the conclusion that if any evolutionary phenomenon appears to be inexplicable on this theory it must be accepted at present merely as one of the facts which in the present state of knowledge seems inexplicable."[25]

Fisher accepts the theory of natural selection by default and renders it virtually unfalsifiable by postulating that any phenomenon that does not seem to support the theory must simply be understood as a reflex of our current deficient state of knowledge. If no phenomena seeming to be contrary to the expectations set up by the theory can be understood as evidence potentially falsifying the theory, Fisher would seem to have violated a fundamental tenet of good science—a good scientific theory must be falsifiable. Rather than engage in an empirical argument here, Fisher advocates Darwin's theory, like Weismann had before him, based on a philosophical premise. Natural selection provides a nonteleological, fully naturalistic explanation of evolution. Other theories at least carry the implication that evolution might be a process directed by some power (natural or supernatural), and the atheist Fisher simply could not abide this possibility. So natural selection wins by default. But the truth of scientific theories cannot be established simply by showing alternative theories to be problematic. It must be established by positive evidence, and this Fisher, like Darwin, largely lacked.

The desire to place the emerging field of biology on a firm naturalistic basis appears to have been Fisher's primary motive for promoting natural selection as the sole driver of evolution. Fisher was acutely aware of the status that physics had attained as the queen of the sciences in his day. This was the era of Albert Einstein, relativity theory, and quantum theory. Physics was rigorous, quantitative, and experimental. Fisher tried to imitate physics by developing a sophisticated mathematical framework for population genetics, even pronouncing a *fundamental theorem of natural selection*—"The rate of increase in fitness of any organism at any time is equal to its genetic variance in fitness at that time"—and

---

24. Ibid., 21.
25. Ibid.

comparing his fundamental theorem to the second law of thermodynamics: "Professor Eddington has recently remarked that 'The law that entropy always increases—the second law of thermodynamics—holds, I think, the supreme position among the laws of nature.' It is not a little instructive that so similar a law should hold the supreme position among the biological sciences."[26] Recall once again that we saw in chapter 1 how Douglas Futuyma and Stephen Stearns celebrated natural selection's role in making biology a science and raising it to the level of chemistry and physics. Recall also Michael Denton's comment that natural selection is really the only truly scientific (naturalistic) understanding of evolutionary mechanism. Fisher saw the importance of natural selection for the creation of a fully scientific biology and so championed it, getting in the bargain a theory that also supported his strongly eugenist proclivities. This is not to say that Fisher did not contribute to our understanding of how gene frequencies change over time in populations, but this is very different from proving that natural selection is the driving force of evolution.

A critique of Fisher's fundamental theorem recently appeared in the *Journal of Mathematical Biology*. William Basener and John Sanford view Fisher's theorem as overly simplified because Fisher excluded the occurrence of mutations from his model, assuming that the net effect of mutations in a population of organisms would be neutral in regard to fitness. That is, negative and positive mutations would cancel each other out over time. Excluding mutations from the model, Fisher believed that natural selection working on genetic variability in the population would always lead to increases in fitness and therefore continuous evolution. But Basener and Sanford show that Fisher's assumption is wrong. Mutations do not have a net-neutral effect on fitness but are more likely to degrade the fitness of the population over time. They write: "As a general rule, the simplifying assumptions as are required for a purely mathematical approach to population genetics force researchers to ignore many variables that tend to reduce the efficiency of selection."[27] Fisher's fundamental theorem is much more important for its role in creating the impression of a rigorous, quantified, fully scientific biology than for its role in accounting for the origin of species. In no sense did Fisher definitively solved Darwin's problem.

26. Ibid., 39.
27. Basener and Sanford, "Fundamental Theorem."

Sewall Wright

The American population geneticist Sewall Wright (1889–1988) enjoyed a long career working on issues related to population genetics and evolution. His initial experience was with animal breeding at the U.S. Department of Agriculture where he worked out some of his own mathematical models (which differed somewhat from Fisher's) to predict how gene frequencies would change over time in a breeding population of livestock. Wright knew from this experience that when animal breeders relied on mass selection in large populations, they made very slow progress toward the creation of their desired trait. But intentional inbreeding in a herd of cattle uncovered hidden variability that could speed up the progress toward the desired trait. Wright theorized that mass selection in large natural populations would be inefficient as well, but if those populations were divided into smaller breeding subgroups, unintentional inbreeding could uncover hidden variability, leading to the development of new traits on which natural selection could act. Fisher thought that evolution would proceed best with mass selection in large populations and considered the effective breeding size of a population to include its complete worldwide distribution. Wright thus differed greatly with Fisher over what breeding population size would be necessary for evolution to occur, an issue they debated for years without ever coming to a resolution.

Wright was also troubled by nonadaptive traits exhibited by many organisms. He took seriously work done by G. C. Robson and O. W. Richards on nonadaptive differentiation among species.[28] Robson had published *The Species Problem* in 1928 to point out how the traits used to distinguish two closely related species from each other often cannot be explained via recourse to greater levels of adaptation, as one would expect on a strictly Darwinian model.[29] Natural selection only preserves what is adaptive, so the traits distinguishing one species from another should show signs of adaptation. But in many cases they don't. Wright thus theorized that in small populations, certain genes could become fixed even if they produced traits that were adaptively neutral by a process

---

28. Richards and Robson had written, "It thus seems the direct utility of specific characters has rarely been proved and it is at any rate unlikely to be common ... Thus the role of Natural Selection in the production of closely allied species, so far as it is known at present, seems to be limited," (Richards and Robson, "Species Problem and Evolution," 384).

29. Robson, *Species Problem*, 216.

called random genetic drift. He believed drift was another mechanism of evolution that acted alongside natural selection, though the importance he placed on genetic drift declined later in his career as the importance of natural selection was championed during the period of the evolutionary synthesis. But here again, Wright disagreed sharply with Fisher, who always held to the sole sufficiency of natural selection.

Wright also differed from Fisher on basic philosophical issues. Fisher was a hardened materialist. But Wright believed that mind was a fundamental aspect of the world that could not be reduced to material explanation. He called this philosophy *monistic panpsychism* and believed it rendered science a limited venture:

> It is the task of science, as a collective human undertaking, to describe from the *external* side, (on which alone agreement is possible), such statistical regularity as there is in the world in which every event has a unique aspect, and to indicate where possible the limits of such description. It is not part of its task to make imaginative interpretations of the internal aspect of reality—what it is like, for example, to be a lion, an ant or an ant hill, a liver cell, or a hydrogen ion.[30]

Wright seems to be proposing an idea similar to that of the French Jesuit priest and paleontologist Pierre Teilhard de Chardin, who spoke of "the within of things" to describe how elementary forms of consciousness must exist at lower levels of organization to account for the reality of the more complex forms of consciousness that appear at higher levels. Of course, Teilhard's view of evolution was also highly teleological and religiously oriented, a place that Wright seems not to have gone. But it is unlikely that Fisher would have ever wondered what it is like to be an ant or a hydrogen ion! In sharp contrast to Fisher, however, Wright warned, "In carrying out this program [of external analysis], the scientist should not, however, deceive himself or others into thinking that he is giving an account of all of reality. The unique inner creative aspect of every event necessarily escapes him."[31] William Provine reports that each time Wright expressed his philosophy, "the response from his colleagues was generally negative."[32] It is not hard to see why. If mind is not reducible to matter, evolution conceived as a purely materialistic process cannot

---

30. Wright, "Gene and Organism," 17 (italics original).
31. Ibid.
32. Provine, *Sewall Wright*, 96.

explain it. If evolutionary theory cannot explain some aspect of living organisms, there is no reason to assume that evolutionary theory itself must be materialist in orientation. But to admit this is to potentially cross the line into religion. This did not seem to bother Wright, who kept his scientific work and philosophical commitments separate, but it definitely would have bothered Fisher and, as we will see, Haldane.

J. B. S. Haldane

J. B. S. Haldane (1892–1964) also had sharp disagreements with his countryman Fisher. Rather than assert his support for natural selection by default and render it virtually unfalsifiable as Fisher had, Haldane admitted that the causes of evolution were debatable,[33] and that criticisms of natural selection were perfectly valid.[34] Among those criticisms was the phenomenon of nonadaptive traits, which Haldane says, "has led many able zoologists and botanists to give up Darwinism."[35] He further recognized that some paleontological evidence seemed to point toward orthogenetic trends in evolution, an idea that he admitted was incompatible with a strictly Darwinian understanding of the evolutionary process.[36] And Haldane admitted to the non-Darwinian possibility of sudden speciation through the process of hybridization (known as *polyploidy*, widely recognized to occur in plants).[37] Haldane continued to hold to a belief (his word) in natural selection as an important cause of evolution, but he recognized in a way that Fisher seems not to have that the evidence for natural selection was less than compelling, and that contrary evidence existed that could not be easily ignored.

What, then, was the basis for Haldane's belief in natural selection? This is not entirely clear. But to Haldane's credit, he is unusually transparent in admitting to the way his own philosophical biases colored his view of evolution. Haldane states a philosophical bias in favor of what he calls *monism*, which he contrasts sharply with any pluralistic philosophy based on an irreducible distinction between mind and matter. Haldane's monism turns out to be a form of materialism that "excludes

---

33. Haldane, *Causes of Evolution*, 3.
34. Ibid., 20.
35. Ibid., 114.
36. Gould, *Structure of Evolutionary Theory*, 514.
37. Løvtrup, *Darwinism*, 308.

[from evolution] the action of a mind or minds higher than that of the evolving individuals, except in so far as such a mind is concerned in the general nature of the universe and its laws, a question too vast to discuss here."[38] In another place, Haldane writes, "materialist thinking in the past has been revolutionary in its effects. It has built up natural science and undermined religion."[39] Haldane's materialism was closely connected to his well-known advocacy of Karl Marx's dialectical materialism and to his support for communist political ideology (he even quotes Lenin in his scientific work[40]). Having laid out his philosophical biases, Haldane then concludes with, "Such are the philosophical prejudices with which I look at evolution."[41] With such a strong philosophical bias in favor of a materialistic view of the world, it is not surprising that Haldane held to a belief in natural selection in spite of the existence of real criticisms of it, which he admitted "are all perfectly valid."[42] How, then, should we assess the work of this influential trio of population geneticists?

Assessing Population Genetics

In no sense can Fisher, Wright, and Haldane be said to have proven the sufficiency of natural selection to explain evolution. They succeeded in developing a rigorous mathematical structure for evolutionary theory and in modeling how gene frequencies might change over time in populations of organisms subject to selection pressures. But this is a far cry from actually demonstrating that natural selection has accounted for large-scale evolutionary change. Moreover, they disagreed over fundamental aspects of how natural selection might work. Fisher argued for large, effective population sizes, whereas Wright opted for small, subdivided populations. Fisher accepted natural selection as the sole cause of evolution, while Wright saw a role for random genetic drift. Haldane could not deny the possibility of orthogenesis and sudden speciation (or saltation). Fisher and Haldane were hardened materialists, while Wright considered the immaterial mind to be an irreducible feature of the universe. Given

---

38. Haldane, *Causes of Evolution*, 158.
39. Dronamraju, ed., *What I Request from Life*, 16.
40. Haldane, *Causes of Evolution*, 155.
41. Ibid., 158.
42. Ibid., 20.

these disagreements, William Provine was forced to conclude his history of the development of population genetics with these words:

> The isolation of all significant parameters affecting population change is difficult under the best conditions with populations in the laboratory. With populations in nature this problem of course greatly increases. Thus with the gap between theoretical models and available observational data so large, population genetics began and continues with a theoretical structure containing obvious internal inconsistencies.[43]

Provine here highlights one of the biggest criticisms leveled at population genetics: how do we know these theoretical mathematical models represent anything actually occurring in nature in the absence of observational data consistent with the models' predictions? Historian Edward Larson has called Haldane a theoretical biologist whose apparent ham-handedness and impatient temperament made it impossible for him to conduct meticulous field or laboratory research in genetics.[44] According to Provine, "Wright and Fisher wrote and argued about the rare desert plant *Oenothera organensis* over a twenty-year period, but Wright never saw one and to my knowledge neither did Fisher."[45] None of the three population geneticists did fieldwork, and all three were acutely aware of the need for population genetics to be based on observations of what actually occurs in nature.

This problem gave rise to one of the most bitter feuds in the history of evolutionary biology. Ernst Mayr, who was not trained as a geneticist, felt that too much emphasis was being placed on the theoretical work of the population geneticists to the detriment of the work of systematists, Mayr's specialty. Mayr coined the disparaging term "beanbag genetics" to describe how population geneticists focused their theoretical work on only one or at most a few genes at a time—like plucking beans from a beanbag—while ignoring the insights of systematists who analyzed the interactions between whole organisms and their environment. Mayr wrote:

> But what, precisely, has been the contribution of this mathematical school to the evolutionary theory, if I may be permitted to ask such a provocative question? . . . Perhaps the main service

---

43. Provine, *Origins of Theoretical Population Genetics*, 178.
44. Larson, *Evolution*, 223.
45. Provine, *Sewall Wright*, 74.

of the mathematical theory was that in a subtle way it changed the mode of thinking about genetic factors and genetic events in evolution without necessarily making any startlingly novel contributions.[46]

Wright was stung by Mayr's condescending attitude toward population genetics and never, according to Provine, missed an opportunity to criticize Mayr in print. And when Mayr organized a conference on the evolutionary synthesis in 1974, he failed to invite Wright because he thought Wright's role in the synthesis had not been pivotal.[47] This feud never subsided even as Mayr and Wright both lived well past the age of ninety while fighting for their rightful place in the evolutionary synthesis. But the issue that divided them is an important one: is evolution best understood at the level of genes and DNA or at the macrolevel of whole organisms? Even today evolutionary scenarios based on a DNA analysis often give strikingly different answers to divergence times between common ancestors than what is documented in the fossil record—estimates sometimes differ by as much as half a billion years![48] The rift between biologists who focus on microlevel analysis and those who focus on macrolevel structures remains.

The prominent evolutionary biologist Francisco Ayala along with his colleague John Avise have written about population genetics:

> These theoretical works had limited impact on the biology of the time for various reasons: they were formulated for the most part in difficult mathematical language; they were almost exclusively theoretical without empirical corroboration—that is, they showed how evolution *could* occur but not that it *did* occur according to the theory—and they were limited in scope.[49]

Ayala and Avise's critique here resonates with John Stuart Mill's reaction to *On the Origin of Species*; Darwin had shown how evolution might have occurred, but he certainly hadn't proven that it did occur in the way he envisioned. Clearly, between the time of Darwin and the period of the population geneticists, empirical support for natural selection had barely

---

46. Mayr, "Where Are We?" 2.

47. Provine, *Sewall Wright*, 482.

48. For example, see Doolittle et al., "Determining Divergence Times"; Wray, "Molecular Evidence"; Valentine, "Fossils, Molecules, and Embryos."

49. Ayala and Avise, eds., *Essential Readings in Evolutionary Biology*, 104 (italics original).

increased, leaving biologists still in the mode of developing speculative scenarios to show how evolution *might* have happened. Of course, the sophisticated mathematical formalism of the population geneticists gave the impression that natural selection had been placed on a firmer scientific footing, but few biologists of the day were even capable of assessing the mathematical work.

As we will see in the next chapter, the grand narrative of Darwinian triumph will hail Theodosius Dobzhansky as the essential founder of the evolutionary synthesis who brought the theoretical work of the population geneticists together with the work of naturalists and systematists to show that the theoretical models did indeed describe what happened in nature. Dobzhansky did enjoy a long working relationship with Sewall Wright, yet it is clear that Dobzhansky was completely flummoxed by Wright's mathematical reasoning. In one noteworthy passage, Dobzhansky said about Wright:

> He has a lot of extremely abstruse, in fact almost esoteric mathematics. Mathematics, incidentally, of a kind which I certainly do not claim to understand. I am not a mathematician at all. My way of reading Sewall Wright's papers, which I still think is perfectly defensible, is to examine the biological assumptions the man is making, and to read the conclusions he arrives at, and hope to goodness that what comes in between is correct. 'Papa knows best' is a reasonable assumption, because if the mathematics were incorrect, some mathematician would have found it out.[50]

Dobzhansky was not alone in struggling to fully understand Wright's papers. Edger Anderson, director of the Missouri Botanical Garden, wrote to Wright in 1938: "Thank you for the letter. Your interesting computations I accept as I do the pleasant sunshine this afternoon: something which I appreciate and value even though I do not understand it."[51] We will have to ask how Dobzhansky could be hailed as the founder of the grand synthesis of evolutionary biology when by his own admission he was not competent to fully assess one side of the supposed synthesis—the mathematical work of the population geneticists.

Despite the valid criticisms leveled at the work of Fisher, Wright, and Haldane, population genetics as a field of inquiry did not die with

50. Dobzhansky, "Oral History Memoir" (cited in Provine, *Sewall Wright*, 346).

51. Letter from Anderson to Wright dated Oct. 14, 1938, (cited in Provine, *Sewall Wright*, 399).

their passing. This field has continued to develop up to the present day. But so have the criticisms. Massimo Pigliucci, for example, views population genetics as an important part of evolutionary theory but believes it is a "gross simplification to see evolution as fundamentally a matter of changes in gene frequencies over time."[52] Pigliucci agrees with Ayala and Avise in characterizing population genetics as a limited theory that deals with relatively simple situations involving just a few alleles (variants of a gene). But organisms are far more complex entities involving the interactions of thousands of genes. However evolutionary theory proceeds in the future, Pigliucci is sure that "we will need a lot more than just population genetic theory to make sense of evolution."[53] And perhaps most insightful of all, Pigliucci writes, "Fisher's dream of providing biology with the equivalent of the second principle of thermodynamics needs, I think, to be set aside."[54] The work of Fisher, Wright, and Haldane in the 1920s and 1930s did not prove that natural selection is the primary mechanism of evolution. But it did give the impression that evolutionary theory had made great advancements given that population genetic theory could be made to resemble the mathematically rigorous science of physics. Fisher's dream to "scientize" biology was likely the main motivation for the development of population genetics in the first place. And there are good reasons why he apparently held this motivation.

Scientizing Biology

On page 22 of Fisher's *The Genetical Theory of Natural Selection*, Fisher frames his discussion with the following epigraph from an 1854 essay by Thomas Huxley: "In the first place it is said—and I take this point first, because the imputation is too frequently admitted by Physiologists themselves—that Biology differs from the Physico-chemical and Mathematical sciences in being 'inexact.'"[55] That Fisher cites this Huxlian observation as the context in which he wants his readers to consider his mathematical approach to evolution speaks volumes about his motivations. Fisher wants to address Huxley's complaint by making biology more "exact." But

---

52. Pigliucci, "Proper Role of Population Genetics," 317.

53. Ibid., 323.

54. Ibid., 321.

55. Cited in Fisher, *Genetical Theory of Natural Selection*, 22. Fisher is quoting from Huxley's 1854 essay "On the Educational Value of the Natural History Sciences."

in his zeal to do so, Fisher might have misrepresented Huxley's point, further evidence that Fisher was more interested in reshaping society's view of the emerging biological sciences than in simply offering support for a scientific theory. Citing this quote in isolation gives the impression that Huxley was complaining about the "inexactness" of biology, a problem Fisher will address with his rigorous mathematical quantifications. But a check of the context shows that Huxley actually made this statement, not to decry the notion that biology is "inexact" in comparison to the physical sciences, but rather to excoriate the physiologists for accepting what Huxley sees as a dangerous stereotype about biology. Huxley actually argues that biology *is* the equal of the physicochemical and mathematical sciences in the exactness of both its methods and its results.

Huxley's argument, and Fisher's seeming misrepresentation of it, notwithstanding, in the early decades of the twentieth century, the emerging field of biology did face two considerable problems. First, it was not seen to be rigorously mathematical by many people, making it look less like a proper science than chemistry and physics. Second, it lacked unity. The life sciences were a hodgepodge of different fields—genetics, systematics, zoology, botany, and paleontology—with nothing to bring them together into a unified whole. The practitioners in these different fields rarely communicated with one another or took notice of the work done outside their own specialization. The biological sciences suffered from a lack of clear identity that resulted in a failure to garner the kind of social status afforded the so-called hard sciences. Interestingly, the frequency of the term *biology* did not outpace the frequency of the term *natural history* in English books until around 1920.[56]

Compounding these two problems was the fact that many did not consider evolutionary theory a legitimate scientific field. According to the noted Harvard physiologist of the 1920s W. J. Crozier, "Evolution is a good topic for the Sunday supplements of newspapers, but isn't science: you can't experiment with two million years."[57] The Rockefeller Foundation provided financial support for research in the emerging fields of molecular biology and the biomedical sciences, but not for research in evolution.[58] Even as late as the 1950s, the National Science Foundation failed to list evolutionary biology as an official category of the life sci-

56. Harrison, *Territories of Science and Religion*, 166.

57. From an unpublished autobiographical manuscript of Ledyard Stebbins, (cited in Smocovitis, *Unifying Biology*, 118).

58. Smocovitis, *Unifying Biology*, 67.

ences.⁵⁹ This point is underscored by the recollection of Duke University botanist Janis Antonovics that his first grant proposal to the NSF in the 1970s to study plant evolution was considered by the systematics panel since no panel existed specifically for the evaluation of proposals in evolutionary biology.⁶⁰ Moreover, philosopher Steve Fuller makes the striking observation that no Nobel Prize has ever been awarded specifically for advancements in evolutionary theory. Nobel Prize winners like Thomas Hunt Morgan and Watson and Crick certainly made discoveries important to evolutionary theory. But their Nobel Prizes were classified in other biological disciplines like genetics or molecular biology. Surprisingly, the great shapers of the modern synthesis of evolutionary theory—Dobzhansky, Mayr, Fisher, Wright, Haldane, William Hamilton, John Maynard Smith—never won a Nobel Prize for their work specifically on evolution.⁶¹ Evolutionary theory has always been suspect as a distinct biological discipline.

Nevertheless, those working in the various fields of biology needed a unifying framework to raise the scientific status of their research, and those interested in evolutionary questions needed to finally divorce their work from its natural-theology roots and establish it as a rigorous science. Fortunately, a single solution existed for both problems: provide the theory of natural selection with a mathematical framework, and then use it as a unifying scaffold for all of biology.

This attempt to unify biology occurred against the backdrop of the larger Unity of Science movement emanating from the logical positivism of the Vienna Circle. Based on a principle pronounced by Ernst Mach, who called for the unification of science to happen as the result of the elimination of metaphysics, the Unity of Science movement called for the unification of all sciences around a strictly materialistic understanding of the natural world. Biologists quickly recognized that "Evolution, purged of unacceptable metaphysical elements, would function as the phenomenon that would make biology an 'autonomous' science." It could "lift biology above the physical sciences at the same time that it 'bound' the fractured biological sciences."⁶² Historian of biology Vassiliki Betty Smocovitis uses the analogy of labor unions to describe how evolution-

---

59. Ibid., 69.
60. Antonovics, "Evolutionary Dys-synthesis," 327.
61. Fuller, *Science vs. Religion?* 132.
62. Smocovitis, *Unifying Biology*, 114.

ary biologists in the 1940s and 1950s unified biology and employed a media campaign to convince other scientists that "evolution was a proper science that could rival physics and chemistry." She writes:

> Rituals of celebration, textbooks, museum exhibits, and televised programs could all be seen as part of a well-planned 'strategy' to 'sell' the science of evolution. In return, evolutionists would then receive patronage and support from government and private agencies, have access to material and other resources, and gain both legitimacy and prestige with respect to other sciences.[63]

This program included the founding of a peer-reviewed journal, *Evolution*, with Ernst Mayr as its first editor, who made clear in his letters to potential contributors "the need to stay away from dangerous ideas."[64] Mayr apparently pointed out to one contributor that "the prestige of evolutionary research has suffered in the past because of too much philosophy and speculation."[65]

It is in this context that we must understand Ronald Fisher's default argument in support of natural selection and his desire to formulate a fundamental theorem of natural selection comparable to the second law of thermodynamics. Similarly, we find Ernst Mayr commenting, "For a physicist, one of the most important parameters of any process is its rate. In the endeavor to make evolutionary biology as similar to the physical sciences as possible, evolutionists have attempted from 1859 on to determine evolutionary rates."[66] Moreover, Julian Huxley (grandson of Thomas Henry) remarked how "The particulate nature of inheritance enables calculations to be made as to the proportions of offspring of different types in different generations after a cross. Like the atomic theory in chemistry, it is the basis of quantitative treatment."[67] Indeed, the quantification of evolution was, according to Smocovitis, "part of a process that would eventually lead to general support for natural selection as the primary mechanism of evolution."[68] And this quantification process was driven by the need to purge biology of metaphysical elements. In the words of

---

63. Ibid., 67.

64. Ruse, *Evolution-Creation Struggle*, 187.

65. Letter from Mayr to G. G. Ferris dated March 28, 1948, (cited in Ruse, *Evolution-Creation Struggle*, 187).

66. Mayr, *One Long Argument*, 158.

67. Huxley, *Evolution*, 47.

68. Smocovitis, *Unifying Biology*, 122.

Julian Huxley, "Darwinism thus reborn is a modified Darwinism, since it must operate with facts unknown to Darwin; but it is still Darwinism in the sense that it aims at giving a naturalistic interpretation of evolution."[69]

With evolution placed on a firmer quantitative footing, this more rigorous scientific theory could be reinterpreted as the unifying framework for all the disparate fields of biology. Ernst Mayr begins his seminal text *Animal Species and Evolution* with, "The theory of evolution is quite rightly called the greatest unifying theory in biology."[70] But the best explanation of exactly how evolution unifies biology comes from Julian Huxley's influential 1942 work *Evolution: The Modern Synthesis*:

> The consideration of evolution thus demands data from the following branches of biology. As regards its historical course, directly from paleontology and indirectly from systematics and biogeography. As regards mechanism, from genetics and cytology, and, since the expression of a gene is important, from studies of development and growth; in addition, systematics may throw light on the types of variation to be found in nature. And as regards biological meaning, from physiology and ecology for the study of adaptation; from mathematics, selection experiments, and, indirectly from paleontology, for the study of survival and extinction. All these are necessary, but none of them alone is sufficient.[71]

With this unification, Huxley sees biology now "coming to rival the unity of older sciences like physics, in which advance in any one branch leads almost at once to advance in all other fields, and theory and experiment march hand-in-hand. As one chief result, there has been a rebirth of Darwinism."[72]

We cannot ignore the role that Darwin's theory played in the "scientizing" of biology in the first half of the twentieth century. Once again recall Douglas Futuyma's 2014 pronouncement that Darwin's theory as one of the most important ideas ever devised *because* of its role in making biology a science. The theory of natural selection, what Michael Denton calls the only truly scientific explanation of evolution, was central to purging biology of its metaphysical baggage, unifying the disparate branches of biology, and advancing the guild interests of the biological sciences by

---

69. Huxley, *Evolution*, 27.
70. Mayr, *Animal Species and Evolution*, 1.
71. Huxley, *Evolution*, 42.
72. Ibid., 26.

putting biology on a par with chemistry and physics as a proper science. But just because a theory is crucial to advancing the guild interests of a scientific discipline does not in any way speak to the truth-value of that theory. Scientific truth is based on evidence, not on guild interests, and evidence for natural selection's ability to account for large-scale evolutionary change was still largely lacking during this process during the early twentieth century of scientizing biology.

Criticisms of this scientizing tendency can still be found today within the biological establishment. Massimo Pigliucci and Jonathan Kaplan have referred to evolutionary biology's insistence on the predictive power of statistical analysis as a kind of "physics envy." One of the hallmarks of physics, they argue, has been its ability to make predictions about the future states of physical systems. "A theory of evolutionary forces, combined with a method for deriving values for those forces that could be plugged into the standard equations of population genetics, seems to imply that a similar project would be possible for biology."[73] But as Pigliucci and Kaplan point out, a historical science like evolutionary biology is quite a bit different from an ahistorical one like physics. "The evolutionary biologist is more akin to Sherlock Holmes than to a theoretical physicist: while generalizations about the behavior of criminals are indeed useful, they will not lead us to find the culprit in any particular case. But of course, that dependence on local conditions is the beauty and essence of both good detective novels and good evolutionary biology."[74] As Smocovitis concludes, Fisher, Wright, and Haldane "were attempting to bring biology up to par with the physical sciences, as they drew on, and modeled after, the repertoire of the physical sciences."[75] At the end of the day, quantifying and unifying biology around the framework of the rigorous mathematical models of the population geneticists had considerably more to do with advancing the guild interests of the developing biological establishment than with answering the question that had so puzzled Darwin: what is the origin of species?

Even those who view the posteclipse recovery of Darwinism as the recovery of a true theory that had lain dormant for a time nevertheless recognize that the theoretical models of the population geneticists by themselves could not empirically establish natural selection as the

---

73. Pigliucci and Kaplan, *Making Sense of Evolution*, 61.
74. Ibid., 111.
75. Smocovitis, *Unifying Biology*, 123.

principal mechanism of evolution. Natural selection could only be finally established by showing that these theoretical models accurately reflected actual processes occurring in nature. The theoretical work of the population geneticists had to be brought into conversation with the work of the systematists and naturalists who studied organisms in the wild. In the grand narrative of Darwinian triumph, the Russian immigrant Theodosius Dobzhansky plays the role of the hero who creates this synthesis and finally establishes the truth of natural selection with the 1937 publication of his landmark *Genetics and the Origin of Species*. According to the grand narrative, Dobzhansky's book laid the foundation for a grand evolutionary synthesis that would play out over the next decade, leading to the final establishment of Darwinian natural selection as the widely accepted explanation of the evolutionary process. After the development of this so-called modern synthesis, the work of evolutionary biologists would simply revolve around detailing in ever greater depth and precision exactly how natural selection (with some help from gene flow and genetic drift) has driven the evolutionary process. The basic Darwinian framework would no longer be challenged. In the vocabulary of Thomas Kuhn, with the paradigm of Darwinian evolution now firmly in place, biologists would set about the work of doing "normal science" under the influence of the paradigm.

But was there truly an evolutionary synthesis, and did biologists acquiesce to the Darwinian paradigm and simply content themselves with the routine matters of normal science, or, as Kuhn called it, puzzle solving? As I will detail over the next two chapters, the grand narrative of Darwinian triumph once again obscures a far more ambiguous but also far more interesting and important historical narrative.

4

Reframing the Modern Synthesis: The Reality of Natural Selection

NO PERIOD IN THE history of evolutionary theory is more fundamental to the grand narrative of Darwinian triumph than the period known as the modern synthesis. The shapers of the grand narrative readily admit that population genetics did not fully establish natural selection as the primary mechanism of evolution. Population geneticists had merely shown how natural selection *might* have driven the evolutionary process. But without extensive fieldwork on organisms in the wild, it was impossible to prove that theoretical models from population genetics represented processes actually occurring in nature. Thus, population genetics had to be brought into conversation with the work of naturalists to see if the theoretical models could be synthesized with data from natural populations. According to the grand narrative, a series of seminal works published over a little more than a decade achieved this synthesis of all the various branches of biology around the concept of Darwinian evolution.

In the words of Ernst Mayr, a leading figure in the synthesis:

> What occurred during the period from 1936 to 1950, when the synthesis took place, was not a scientific revolution; rather it was a unification of a previously badly split field. The evolutionary synthesis is important because it has taught us how such a unification may take place: not so much by any revolutionary new concepts as by a process of house cleaning, by the final rejection of various erroneous theories and beliefs that had been responsible for the previous dissension. Among the constructive

achievements of the synthesis was the finding of a common language among the participating fields and a clarification of many aspects of evolution and its underlying concepts.[1]

By Mayr's own admission, the triumph of Darwinism resulting from the synthesis was not based on the development of new theories or the accumulation of new evidence, but rather resulted from "house cleaning" and the "final rejection of various erroneous theories and beliefs." In other words, evolutionary biologists circled the wagons and marginalized theories not consistent with the growing consensus around natural selection—a consensus that, as I argued in the previous chapter, was motivated at least in part by an attempt to unify biology and make it a naturalistic science. This process of "house cleaning" had the effect of enshrining a series of seminal works that supported the growing Darwinian consensus as the authoritative canon of evolutionary biology.

This canon begins with Theodosius Dobzhansky's *Genetics and the Origin of Species* (1937), credited with demonstrating consistency between population genetics and studies of natural populations; and it includes Julian Huxley's *Evolution: The Modern Synthesis* (1942), zoologist Ernst Mayr's *Systematics and the Origin of Species* (1942), paleontologist George Gaylord Simpson's *Tempo and Mode in Evolution* (1944), and zoologist Bernhard Rensch's *Neuere Probleme der Abstammungslehre* (1947)—later translated into English as *Evolution above the Species Level*—and botanist George Ledyard Stebbins's *Variation and Evolution in Plants* (1950). Mayr calls these authors the "architects of the evolutionary synthesis."[2]

But did such a grand synthesis really occur, and if so did it succeed in placing natural selection on a secure foundation as the clear mechanism of evolutionary change? More specifically, can natural selection be shown to occur in nature, and if it can, is it possible to demonstrate that natural selection accounts for large-scale evolutionary change? To answer these questions, I will critically engage several works of the synthesis. In this chapter, the focus will be on the works of Dobzhansky and Simpson, which will lead to a consideration of later works on natural selection by John Endler and Graham Bell along with a case study on the phenomenon of industrial melanism. The question here is, has the modern synthesis established the reality of natural selection (microevolution)

---

1. Mayr, *One Long Argument*, 134.
2. Ibid.

in the wild? In the next chapter, I will consider Bernhard Rensch's work along with an analysis of the chief noncanonical work produced during the period of the synthesis, Richard Goldschmidt's *The Material Basis of Evolution* (1940). The focus there will be on the question, can natural selection account for large-scale evolutionary change (macroevolution)?

Theodosius Dobzhansky

Born in the Ukraine in 1900, Theodosius Dobzhansky came to the United States in 1927 on a fellowship to work in the laboratory of Thomas Hunt Morgan at Columbia University. Dobzhansky had studied variation in beetles and fruit flies in Russia and was interested in the work on fruit fly genetics being conducted by Morgan at Columbia. Once in America, he never returned home but enjoyed a long career teaching at the California Institute of Technology, Columbia University, and the Rockefeller Institute in New York City. He died in 1975. As the author of *Genetics and the Origin of Species*, Dobzhansky and his work occupy a place in the development of modern evolutionary theory that is hard to overstate. But what exactly did he think he had accomplished in this seminal work? Given the status of his book within the canon and grand narrative of Darwinian evolution, it is quite surprising how often Dobzhansky downplays the strength of the evidence for natural selection. One would expect a more confident tone in a book widely touted as the most important work in evolutionary biology after *On the Origin of Species*.

Dobzhansky begins by acknowledging the "species problem" earlier articulated by Robson and Richards, observing that discontinuity between species is a fundamental characteristic of the organic world.[3] Though species are supposed to have evolved, according to Darwin, in a slow, gradual manner, the natural world is not populated today by a plethora of intermediate forms, as one might expect, but by clearly demarcated groups called species. Dobzhansky then continues with a truly striking admission:

> Among the present generation no informed person entertains any doubt of the validity of the evolution theory in the sense that evolution has occurred, and yet nobody is audacious enough to believe himself in possession of the knowledge of the actual mechanisms of evolution. Evolution as an historical process is

3. Dobzhansky, *Genetics and the Origin of Species*, 4.

established as thoroughly as science can establish a fact witnessed by no human eye ... But the understanding of causes which may have brought about this evolution, and which can bring about its continuation in the future, is still in its infancy.[4]

Right from the beginning, Dobzhansky alerts us not to read *Genetics and the Origin of Species* as a convincing proof of the reality of natural selection. Though Mayr may have been "audacious enough to believe himself in possession of the knowledge of the actual mechanisms of evolution," Dobzhansky certainly does not think the evidence warrants such confidence.

Dobzhansky continues to express doubts about other fundamental aspects of evolutionary theory. On the exact nature of the mutational and chromosomal changes that lead to the variation on which natural selection acts, Dobzhansky pleads ignorance, believing he is in no better position than Darwin was on this point. Further, he realizes that even if a complete knowledge of the physiological mechanisms causing gene and chromosome mutation could be attained, it would not necessarily tell us much about evolution. Variations are the raw building material for evolution, "but the presence of an unlimited supply of materials does not in itself give assurance that a building is going to be constructed."[5] Dobzhansky recognizes that the origin of hereditary variation is an entirely different problem from the origin of species. An understanding of the former is no guarantee of an understanding of the latter. On the latter, Dobzhansky does realize that population geneticists had produced rigorously quantitative models to demonstrate how variations, no matter how they occurred, could be propagated through a population of organisms. Nevertheless he writes:

> Only in recent years, a number of investigators, among whom Professor Sewall Wright of Chicago should be mentioned most prominently, have undertaken a mathematical analysis of these processes, deducing their regularities from the known properties of the Mendelian mechanism of inheritance. The experimental work that should test these mathematical deductions is still in the future, and the data that are necessary for the determination of even the most important constants in this field are wholly lacking.[6]

4. Ibid., 8.
5. Ibid., 119.
6. Ibid., 120.

This is a surprising admission considering that the grand narrative views Dobzhansky's book as having synthesized the theoretical models of the population geneticists with experimental work on natural populations. For Dobzhansky, experimental corroboration of the theoretical models still lay in the future. Even important constants in those models could not be calculated. The models thus remained speculative.

As we saw in the previous chapter, Sewall Wright and Ronald Fisher engaged in a long and deep disagreement over the issue of the effective population size most consistent with Darwinian evolution. Fisher argued for large, randomly breeding populations, whereas Wright thought evolution would proceed best in smaller, isolated subpopulations. Dobzhansky agreed with Wright, but admitted: "the discussion of the population sizes can not but be restricted to bare generalities. We shall have to be content if we can assemble some data from which inferences may be drawn that could serve as provisional working hypotheses for future investigations."[7] Dobzhansky relegated Wright's preference for smaller semi-isolated subpopulations to the status of a "fruitful hypothesis."[8] Once again, he considered empirical confirmation of a fundamental aspect of evolutionary theory a future possibility, but not a present reality, when he was writing.

Since evolutionary change is not noticeable in most species within a human lifetime, Dobzhansky considered the idea of natural selection as the driving force of evolution to be "at best no more than a very probable inference."[9] Of course, even if phenotypic change (change to an organism's form, function, and/or behavior) is not noticeable on human timescales, in some circumstances a change in the genetic composition of a population of organisms can be observed within historical time. Would this not indirectly strengthen the case for natural selection? Even if we do not directly observe phenotypic change, the ability to document genetic change might lead to the conclusion that natural selection is occurring and would inevitably produce phenotypic change given enough time. But even in this case, Dobzhansky demurs: "The occurrence of such changes [genetic changes] is *per se* not a sufficient proof that the transformation is due to selection. To make such an assumption would mean taking for granted what one pretends to be trying to prove."[10] Dobzhansky

7. Ibid., 139.
8. Ibid., 148.
9. Ibid., 151.
10. Ibid., 159.

well understood just how difficult it is to prove not only the existence of natural selection itself but also natural selection as the process primarily responsible for large-scale evolutionary change. One can observe small-scale change in a population of organisms (e.g., changes in beak size of finches in the Galápagos Islands or the shift in frequency of color morphs in a species of moth or butterfly), and one can assume that these changes were driven by natural selection. But it is much more difficult to actually document the greater reproductive fecundity of organisms bearing a certain trait without being able to track the life histories of thousands or millions of individual organisms in the wild over many generations and to document the life histories of each of their progeny. Not surprisingly then, Dobzhansky states outright, "The inadequacy of the experimental foundations of the theory of natural selection must be admitted, I believe, by its followers as well as by its opponents."[11]

It is instructive to compare Dobzhansky's circumspect attitude toward natural selection with the more confident pronouncements of Ernst Mayr. In *Systematics and the Origin of Species*, Mayr considers the selective value of color variation in organisms. "If the color of an animal agrees with the background on which it lives, the survival value of this adaptation is obvious."[12] Mayr then illustrates this principle by considering the example of Palearctic larks, which live in areas with scanty vegetation, making them easy targets for predators. Different races of the lark live on different kinds of soil and appear to develop coloring that matches the soil background on which they live, thus hiding them from predators. In the Middle East, Mayr says, a dark-colored subspecies of the lark was found living on a dark lava flow. This subspecies was sandwiched between two lighter varieties living on sandy soil. To Mayr, "It is obvious that selection by predators must have played an important role in the evolution of these localized color races."[13] But a moment's thought raises an important question: If a lighter race of larks became dark after moving to the lava flow, how did the initial colonists to the lava flow escape predation when they would have still been lighter in color? Since natural selection works slowly and gradually, it would have taken a number of generations before the genetic mutation conferring a protective dark color spread through the population if it occurred at all. In the

---

11. Ibid., 176.
12. Mayr, *Systematics and the Origin of Species*, 86.
13. Ibid., 87.

meantime, the lighter-colored larks would have been extremely vulnerable to predation and could easily have been driven to local extinction. This is the problem raised earlier by Weismann: the selective advantage of the initial stages of variation. Weismann was forced to simply assume a selective advantage for the initial stages of variation because he had already assumed the truth of natural selection for ideological reasons. There is no reason to think, however, that Mayr's confidence on this point is based on any firmer ground than Weismann's. Dobzhansky recognizes in a way that Mayr ignores how natural selection as an explanation for some observed phenomenon could sound logical and convincing in the abstract. But without direct experimental confirmation, it remains at best "no more than a probable inference."[14]

For a book supposed to have ushered in a grand synthesis of theoretical population genetics with data from natural populations in order to demonstrate the reality of natural selection as the primary mechanism of evolution, *Genetics and the Origin of Species* is unexpectedly filled with hesitations, caveats, and candid admissions of the difficulties encountered in actually documenting the action of natural selection. It also frankly admits to the many lacunas in our knowledge. It does not appear that Dobzhansky thought he had definitively solved the problem of the origin of species. Far from it. Even twenty-two years later he still could say that the evolutionary synthesis was unfinished, and that "radical changes and major upsets are not only possible but almost certain to occur."[15] The idea that Dobzhansky initiated a grand evolutionary synthesis appears to be more the function of the mythmaking of the grand narrative of Darwinian triumph than a reporting of actual history. Just how confident was Dobzhansky in the role of natural selection in evolution? It is hard to say.

Though Dobzhansky seems in many ways to have been in step with the emerging Darwinian orthodoxy of his time, other aspects of his life and career seem to betray a certain ambivalence. Dobzhansky was a deeply religious man throughout his life, a devout Russian Orthodox Christian. He had an abiding interest in the larger implications of evolutionary theory for fields such as sociology and philosophy, especially for questions related to human societies and ethics. He believed the origin of life and the evolution of the human mind were the evolutionary problems most in need of explanation since both involved a major qualitative

14. Dobzhansky, *Genetics*, 151.
15. Dobzhansky, Foreword to *Evolution above the Species Level*, by Bernhard Rensch, v.

increase in complexity.[16] His realization of the complexity required for the origin of life and the development of the human mind lay behind Dobzhansky's interest in the work of the French Jesuit and paleontologist Pierre Teilhard de Chardin. Teilhard is known for developing a teleological and ultimately religious understanding of the evolutionary process. He tried to account for how evolution became intelligent and reflective so that the human mind evolved toward union with God. Dobzhansky, surprisingly, served as president of the North American Teilhard Society for the year 1969, and in a famous essay published late in life, he called Teilhard "one of the great thinkers of our age."[17] In order to justify his own synthesis of Christian faith with evolutionary theory, Dobzhansky cited Teilhard as an example of a devout Christian who nevertheless accepted evolution. But how could he ignore the highly teleological nature of Teilhard's view of evolution? Could Dobzhansky have served as president of the Teilhard Society and not have in some way been influenced by the profoundly non-Darwinian character of Teilhard's work? This question sharpens when we consider how Teilhard's work was viewed by other advocates of Darwinism.

Nobel Prize-winning biologist Peter Medawar, for example, called Teilhard's book *The Phenomenon of Man*

> a book widely held to be of the utmost profundity and significance; it created something like a sensation upon its publication in France, and some reviewers hereabouts called it the Book of the Year—one, The Book of the Century. Yet the greater part of it, I shall show, is nonsense, tricked out with a variety of metaphysical conceits, and its author can be excused of dishonesty only on the grounds that before deceiving others he has taken great pains to deceive himself. The *Phenomenon of Man* cannot be read without a feeling of suffocation, a grasping and flailing around for sense.[18]

More recently Daniel Dennett has written of Teilhard:

> It is fair to say that in the years since his work was published, it has become clear to the point of unanimity among scientists that Teilhard offered nothing serious in the way of an alternative

---

16. Taylor, "Dobzhansky, Artificial Life, and the 'Larger Questions' of Evolution," 168.

17. Dobzhansky, "Nothing in Biology," 129.

18. Medawar, "Critical Notice: *The Phenomenon of Man*," 99.

to orthodoxy; the ideas that were particularly his were confused, and the rest was just bombastic redescription of orthodoxy.[19]

Dennett charges Teilhard with denying Darwin's fundamental idea: "that evolution is a mindless, purposeless, algorithmic process,"[20] and argues that the esteem with which nonscientists continue to hold Teilhard's book is simply a testimony to their "depth of loathing of Darwin's dangerous idea, a loathing so great that it will excuse any illogicality and tolerate any opacity in what purports to be an argument."[21] By Dennett's definition, then, Dobzhansky would count as a nonscientist! For Dobzhansky manifestly did hold Teilhard in high esteem.

For his part, Medawar confesses to being able to read Teilhard's book only "with real distress, even despair."[22] But not so Dobzhansky, who saw Teilhard's work as expressing a legitimate fusion of scientific and religious thought issuing from "one of the great thinkers of our age." Dobzhansky even goes so far in same the article in which he extols Teilhard as to refer to himself as both "a creationist *and* an evolutionist," because "Evolution is God's, or Nature's method of Creation."[23] And this was in 1973, after the development of modern creation science movements in the 1960s. That Dobzhansky would own the title "creationist" in such a fraught social context is truly surprising. I will consider Dobzhansky's form of theistic evolution in more detail in chapter 8. For now, we have to at least raise the question of Dobzhansky's full commitment to a nondirected Darwinian view of evolution. Whatever his real views may have been, Dobzhansky is the man the grand narrative enshrines as the hero of its story of the triumph of Darwinian evolution. He appears to have been a reluctant hero.

George Gaylord Simpson

Born in 1902, George Gaylord Simpson (d. 1984) became one of the greatest of all American paleontologists and enjoyed a long career teaching at Columbia University, Harvard University, and the University of Arizona. In 1944, he published *Tempo and Mode in Evolution*, a work credited with demonstrating how the fossil record could be understood

19. Dennett, *Darwin's Dangerous Idea*, 320.
20. Ibid.
21. Ibid.
22. Medawar, "Critical Notice: *The Phenomenon of Man*," 106.
23. Dobzhansky, "Nothing in Biology," 127.

as consistent with the process of natural selection. This was no small feat, considering that paleontologists were among the most reluctant members of the biological community to accept Darwinian evolution. Stephen Jay Gould reports, "When Dobzhansky heralded the modern synthesis in 1937, I do not think that a single prominent paleontologist had recorded himself as a consistent Darwinian."[24] Perhaps this is because as Julian Huxley remarked, "Paleontological study of the *course* of evolution cannot be decisive in terms of the *method* of evolution."[25] Fossils don't speak for themselves. They must be interpreted. Just because one is able to line up a series of fossils showing clear evolutionary development, the fossil record itself cannot show *how* that evolution occurred. So on what grounds has Simpson's seminal text become part of the Darwinian canon? Let us consider some aspects of his argument.

The first interesting point made by Simpson concerns evolutionary rates. On a Darwinian model, one would expect small animals to evolve more quickly than larger ones because smaller animals have shorter generation times. If evolution is a process that occurs in populations, then the more quickly a population genetically turns over, the more quickly new adaptive traits can become established. Simpson observes, however, that when we look at the fossil record of mammals who entered South America from North America one to two million years ago, there seems to be no difference in evolutionary rates between larger and smaller animals. Simpson continues:

> Much of the evidence of this sort is so vague and unsatisfactory that its interpretation is almost entirely subjective, and the same facts may be used to reach diametrically opposed conclusions. Thus the great endemism in faunas of isolated Pacific islands has been cited as exemplifying slow evolution on very old islands and also as proof that evolution has there occurred very rapidly.[26]

Simpson here echoes a point made by Thomas Kuhn in *The Structure of Scientific Revolutions*: "Philosophers of science have repeatedly demonstrated that more than one theoretical construction can always be placed upon a given collection of data."[27] The highly interpretive nature of fos-

---

24. Gould, "G. G. Simpson," 154.
25. Huxley, *Evolution*, 38 (italics original).
26. Simpson, *Tempo and Mode in Evolution*, 20.
27. Kuhn, *Structure of Scientific Revolutions*, 76.

sil evidence makes it particularly vulnerable to subjective bias. Simpson provided another example of this in a later work:

> Some paleontologists have been so impressed by the frequent trend for animals to become larger as time goes on that they have tried to work it the other way around. If they find, say, a Pleistocene bison that is somewhat larger than a Recent bison ... they conclude that it is not ancestral to later bison *because* it is larger. You can establish any "rule" you like if you start with the rule and then interpret the evidence accordingly.[28]

On a Darwinian model, one would expect some correlation between the rate at which mutations occur and the rate of evolutionary change. Faster mutation rates would seem to provide more opportunities for adaptive traits to emerge, thus fueling a faster rate of evolution. But Simpson responds: "With regard to the question of size and nature of mutations ... it is quite impossible to count mutations that occur in the history of any group of animals and ... there is therefore no direct observational approach to the study of mutation rates during that history."[29] Given the impossibility of counting mutations, "Any attempt to judge whether observed mutation rates are consistent with observed evolutionary rates involves so many untestable hypotheses and subjective estimates that little sober scientific value can be claimed for it."[30] Recall that in the previous chapter we saw Ernst Mayr discussing how concern with evolutionary rates among biologists was part of how biology was being made into a science. No biologist is in a better position than a paleontologist to assess evolutionary rates, but in Simpson's view, the subjective nature of paleontological interpretation undermines this quantitative aspect of evolutionary theory.

Despite these caveats, Simpson makes it clear he is still a proponent of gradual Darwinian evolution. He does not accept the possibility of sudden, large-scale evolutionary change because multiple coordinated mutations would be necessary to bring such changes about. He calculates that five mutations occurring in an individual organism that is part of a population of 100 million individuals with a generation time of one day would be expected to occur only once every 274 billion years, about 100 times the age of the earth. Nothing short of divine intervention could

---

28. Simpson, *Meaning of Evolution*, 137.
29. Simpson, *Tempo and Mode in Evolution*, 44.
30. Ibid., 45.

produce this, according to Simpson, so saltational evolution cannot have occurred.[31] If mutations occur randomly as Simpson assumes, his calculations appear accurate. But recall that both Mivart and Bateson argued that multiple coordinated mutations *must* occur for organisms to maintain integrated function during the evolutionary process. An isolated mutation conferring on an animal slightly longer legs would not necessarily allow that animal to run faster and escape predators more easily if it was not coordinated with simultaneous changes conferring greater cardiovascular and metabolic capacity to support faster running. Longer legs absent these other changes might actually make an animal less coordinated and more vulnerable to predation. We will take up the pressing question of the randomness of mutation in chapter 6.

Next, Simpson takes on the issue of micro- versus macroevolution. *Microevolution* refers to small-scale changes that take place in a population of organisms—changes that do not involve major alterations to the form of an organism: changes like antibiotic resistance in bacteria, pesticide resistance in insects, and shifting proportions of color morphs in insects or coat color in mammals. Given enough time, can these small-scale changes accumulate to a degree large enough to account for large-scale, macroevolutionary change, like the emergence of whole new classes of organisms? In other words, is macroevolution just microevolution writ large, or is it a whole new thing requiring its own unique explanation? In Simpson's view, if micro- and macroevolution prove to be two different processes, then "the innumerable studies of micro-evolution would become relatively unimportant and would have minor value in the study of evolution as a whole."[32] That this might be the case is supported by a fact that Simpson recognizes: the majority of species and genera appear in the fossil record suddenly, differing sharply from earlier groups of organisms. In most cases we don't see smooth gradations of intermediate forms. And, Simpson notes, "this appearance of discontinuity becomes more common the higher the level, until it is virtually universal as regards orders and all higher steps in the taxonomic hierarchy."[33]

As we will see in the next chapter, the large gaps in the fossil record puzzled Darwin. He well knew that the fossil evidence at hand was contrary to his theory of gradual, incremental change. But he believed (or

31. Ibid., 54.
32. Ibid., 97.
33. Ibid., 99.

at least hoped) that future fossil discoveries would fill in the gaps. Since only a small fraction of all the organisms that have ever lived fossilize, clearly the fossil record is highly deficient. Lack of intermediate fossils could simply be the result of poor preservation. Yet the fossil record had certainly improved by Simpson's time, but he still concluded that lack of preservational opportunities could not explain the remaining gaps in the fossil record for many invertebrates and virtually all vertebrates.[34] On what grounds then can Simpson be said to have contributed to the triumph of natural selection? In an interesting passage Simpson considers the discontinuity between the canid group (dogs) and the felid group (cats).

> I am fully confident that animals of completely intermediate structure between primitive true felids and primitive true canids did exist, but the fact that they did not persist and were evidently rare and local (no remains are known) is in itself evidence that there is a natural equivalent of discontinuity between the felid and the canid adaptive zones.[35]

Simpson is confident in the existence of intermediate forms that do not appear in the fossil record because he surmises they must have been rare and local. The discontinuity between modern dogs and cats came about due to discontinuity in the environmental contexts to which their ancestors adapted. But this is supposition, not empirical demonstration. As Simpson himself noted, "You can establish any 'rule' you like if you start with the rule and then interpret the evidence accordingly." Simpson assumes the truth of natural selection, so he must assume the existence of fossil forms for which there is no record.

Still, Simpson was aware of fossil evidence seeming to document relatively rapid evolutionary change that did not appear consistent with a Darwinian process. So he introduced the concept of quantum evolution to describe this, borrowing the term *quantum* from physics, where a quantum is the amount of energy required to cause an electron to leap from one energy state to another. Electrons do not move smoothly between energy states. They stay at one state until absorbing just enough energy to literally make a "quantum leap" to the higher state. Likewise, Simpson surmised that in some circumstances, some organisms could make such a quantum leap as well. He writes, "It is the present thesis

34. Ibid., 116.
35. Ibid., 190.

that quantum evolution also occurs on a larger scale and in clear distinction from any usual phase of speciation or phyletic evolution."[36] Simpson thought of quantum evolution as a new type of macroevolutionary process clearly distinct from normal Darwinian gradualism. He considered it the dominant process in the evolution of higher taxonomic categories like families, orders, and classes. It could help explain "the mystery that hovers over the origins of such major groups."[37]

Talk of an evolutionary process distinct from Darwinian natural selection and of evoking mystery did not sit well with the growing voices of Darwinian orthodoxy. It is not surprising then that when writing a thoroughly revised version of *Tempo and Mode* nine years after the first edition, Simpson backed away from quantum evolution, saying it should not be understood as "a different sort of evolution, or even a distinctly different element of the total phylogenetic pattern. It is a special, more or less extreme and limiting case, of phyletic evolution."[38] But in 1944, Simpson *had* called quantum evolution a process distinct from normal phyletic (gradual) evolution. Why the change? As the evolutionary synthesis progressed into the 1950s, the synthesis hardened around Darwinism, making it more and more difficult for any biologist mindful of their reputation to stray off the Darwinian reservation, even with good empirical reasons to do so. In Simpson's case, there was good empirical evidence to posit the non-Darwinian idea of quantum evolution in 1944, but pressure to conform to the growing Darwinian orthodoxy forced Simpson to back off by 1953. According to Stephen Jay Gould, quantum evolution had by 1953 become what Simpson had explicitly denied before—"merely a name for phyletic evolution when it proceeds at its most rapid rates, a style of evolution differing only in degree from the leisurely, gradual transformation of populations."[39] Darwinism had become a dogma to which biologists felt compelled to assent.

So how do we assess Simpson's overall contribution to the synthesis? His assessment of the paleontological data clearly did not constitute proof of natural selection. According to Gould, Simpson made a consistency argument "based on extrapolating the processes operating within populations to encompass all the phenomena of macroevolution. It worked,

---

36. Ibid., 206.
37. Ibid.
38. Simpson, *Major Features of Evolution*, 389.
39. Gould, "G. G. Simpson," 168.

but it was only a consistency argument, not a proof."[40] Even Ernst Mayr had to agree. Simpson's work (along with work by Rensch, Julian Huxley, and even Mayr himself) "did not prove that macroevolution is governed by variation and selection, but simply showed that macroevolutionary phenomena can be explained in terms of the neo-Darwinian theory."[41] Of course, if one begins with the assumption that natural selection is the driving force of evolution, one can easily interpret fossil evidence through the lens of this assumption and demonstrate consistency. But Simpson had no more than Dobzhansky established natural selection in any kind of empirical sense. Real evidence for natural selection was still missing.

At this point, before considering further the work of the synthesis, it will be helpful to pause and consider two later texts that deal specifically with natural selection—John Endler's *Natural Selection in the Wild* (1986) and Graham Bell's *Selection: The Mechanism of Evolution* (1997). Widely considered seminal works on natural selection, did they place natural selection on firmer empirical ground?

Studies on Natural Selection

Despite being a bit dated, John Endler's *Natural Selection in the Wild* is still considered one of the most important works ever published on the phenomenon of natural selection. Endler (b. 1947) demonstrates the role of natural selection in wild populations and so is credited with providing the evidence to show that theoretical work on natural selection does indeed describe what happens in nature. Indeed, Endler confidently concludes, "Natural selection is common enough in natural populations to have been detected in a wide variety of organisms."[42] How does he know this? Central to Endler's book is a long chart summarizing the details of 154 peer-reviewed studies purporting to demonstrate the action of natural selection in the wild. For each study, Endler includes a note stating whether the selective agent is certain, questionable, or unknown; a *selective agent* is whatever environmental realities determine whether organisms with certain traits survive and reproduce more than organisms lacking these traits. In sixty-nine of the 154 studies, Endler considers the selective agent to be questionable or unknown. Can one really be said to

---

40. Ibid., 169.
41. Mayr, "Role of Systematics," 134.
42. Endler, *Natural Selection in the Wild*, 245.

have demonstrated the action of natural selection in a wild population if one is not sure of the identity of the selective agent in so many studies?

Endler's confidence in the ubiquity of natural selection in the wild seems betrayed by his discussion of the many difficulties one encounters in trying to demonstrate the action of natural selection. According to Endler, "There are many difficulties in detecting natural selection, and these may be divided into three groups according to their effects: (1) lack of detection of natural selection when it exists; (2) apparent detection when it does not exist; and (3) misleading detection of natural selection."[43] Endler considers these difficulties to be surmountable with careful and intensive fieldwork, but his comment that "The literature contains a very large number of studies on natural selection. They vary greatly in quality." suggests that in many cases, the difficulties he enumerates have not been surmounted. This is reinforced by Endler's lists of what he calls "major gaps" in our understanding of natural selection. These gaps are revealed by questions like this: Why does natural selection occur? How does natural selection occur? What traits are most likely to be affected by natural selection? And, what is the effect of simultaneous natural selection on many traits?[44] If Endler truly considers fundamental questions about how and why natural selection occurs to constitute "major gaps" in our understanding, it is unclear what the basis is for his confident pronouncement that "natural selection is common enough in natural populations to have been detected in a wide variety of organisms." Strikingly, Endler seems to undermine his own conclusion with the comment: "There are few cases in which it is known why natural selection occurs ... A strong demonstration of natural selection combined with a lack of knowledge of its reasons and mechanisms is no better than alchemy."[45]

Similar problems arise in Graham Bell's highly technical monograph *Selection: The Mechanism of Evolution*. Like Endler, Bell begins with a confident pronouncement, "Living complexity cannot be explained except through selection, and does not require any other category of explanation whatsoever."[46] But then we are told: "This book is about selection. It is not about any of a host of related topics. It is not primarily about evolution ... In particular, it is not about adaptation. This is because I

---

43. Ibid., 125.
44. Ibid., 247.
45. Ibid., 164.
46. Bell, *Selection*, xviii.

am more concerned to explain how selection can act, in general, than to describe how it has acted, in particular cases."[47] It is hard to understand how a book that is not primarily about evolution can purport to show that selection *is* the mechanism of evolution. By Bell's own admission, his book is theoretical and should not be expected to provide the missing empirical evidence for selection. But even granting this, the theoretical work itself is problematic on at least two counts.

First, consider the following passage:

> Populations that are evolving in similar environments come to differ because evolution through selection has no foresight: it does not provide any way in which a population can map out the best route towards a predefined goal . . . The route taken by a population will instead be contingent: it will depend on the fortuitous occurrence of variation on which selection can act, and the enormous range of possible variation means that different variants are likely to arise at different times in different populations. Each population is likely to evolve by a different route and arrive at a different destination.[48]

Bell's logic here is impeccable. Darwin's theory of natural selection does predict that the natural world should display a high degree of evolutionary divergence. But as we saw in chapter 2, it doesn't. St. George Jackson Mivart made the reality of convergent evolution one of his main criticisms of Darwin, and convergent evolution has continued to be a well-recognized phenomenon to the present day. It forms the heart of Simon Conway Morris's work, and even Ernst Mayr marveled at the reality of convergence and was forced to introduce the teleological metaphor of the engineer to account for it. Bell's theory is out of step with what we actually observe in nature. This incongruence undermines his attempt to show that selection is the mechanism of evolution.

The second theoretical problem with Bell's work is a bit more involved. Given the low probability of multiple coordinated mutations occurring randomly, Bell must account for how a Darwinian process can produce a series of small, sequential mutations that maintain function in an organism at every step in the sequence such that these small-scale changes can accumulate over time to produce new, complex structures. Bell uses the example of the hemoglobin molecule, a complex protein

---

47. Ibid., xx.
48. Ibid., 19.

molecule consisting of a chain of 128 amino acids in which each amino acid must be in the proper place in the chain for the molecule to function normally. Bell asks: What are the odds that this particular sequence of amino acids could arise by chance so it could be selected? His answer? About one in 10^{200}!

> If the surface of the Earth were filled a metre deep with protein molecules 128 amino-acids long; if all the molecules had different sequences; if every molecule changed its sequence at random every second, without the same sequence ever recurring; and if this process had been going on since the origin of the Earth, nearly five billion years ago—then it is still exceedingly improbable that the hemoglobin sequence would yet have been generated.[49]

To get around this astronomical improbability, Bell employs a word-game analogy to illustrate how natural selection actually works in his view.

Let's begin, Bell says, with a word like *word* and try to turn it into the word *gene* by randomly changing one letter at a time. Bell calculates that if we change one letter at random every second, we could expect to transform *word* into *gene* in about three days' time. Obviously, the time needed for such a transformation would inflate quickly with a much longer word, let alone a full sentence or a 128-amino-acid chain. For Bell, "Evolution cannot work like this. It does not."[50]

To demonstrate how evolution does work, Bell says we must add two additional rules to our word game. First, only letter substitutions that give rise to meaningful words will be allowed. This is fine since in natural selection only functional variations will be preserved. But the second rule is more problematic. Any variants, Bell says, that contain a letter substitution closer to *gene* will be preserved and passed on. Others will be discarded. "There is only one variant that by changing a single letter, increases the resemblance of WORD to GENE, which is WORE."[51] This will replace all other sequences. Next, *wore* can give rise to either *were* or *gore*. *Were*, he says, is a blind alley, so *gore* will be selected. *Gore* then can be turned easily into *gone*, with one final step leading to the target word, *gene*. With these two additional rules in place, the transformation of *word* into *gene* proceeds much more quickly. But Bell's second additional rule

49. Ibid., 20.
50. Ibid., 23.
51. Ibid.

completely undermines his view that natural selection has no end goal or final purpose. In the word-game analogy, Bell has introduced a final target for the sequence of random-letter substitutions. Each individual variation will be compared to the final product, *gene*, to determine its "fitness." But this is exactly what Bell has said natural selection *cannot* do. Individual substitutions of amino acids do not know in advance what the "correct" sequence will be to produce hemoglobin. Bell's analogy utterly fails to solve the intractable problem of how a process of individual random mutations acted on by selection can build up new complex structures even over enormous spans of time. The combinatorial possibilities are just too large.

This is often referred to as the problem of combinatorial inflation and has been noted by mathematicians and engineers, though it is often ignored by evolutionary biologists. It was first brought to light at a well-known conference held at the Wistar Institute in Philadelphia in 1966. In the conference proceedings, Murray Eden, an electrical engineer from the Massachusetts Institute of Technology, commented on the unimaginably vast number of possible amino acid combinations in an average protein molecule and concluded that either functionally useful proteins are common in this combinatorial space, or there must be regular paths cut through this space so that not all combinatorial possibilities can be realized. He sides with the second option: "I would conjecture that what one might call 'genetic grammaticality' has a deterministic explanation and does not owe its stability to selection pressure acting on random variation."[52] As I will discuss in more detail in chapter 7, this problem of combinatorial inflation has become a foundational concept in the development of ID theory. It remains a strong challenge to Darwinism.

We return now to Bell: he comments that "the exuberant diversity of dogs is a striking testimonial to the power of selection to direct adaptive change far beyond the limits of the original population within a few hundred generations."[53] Once again, Bell ignores the crucial difference between the artificial selection practiced by dog breeders and natural selection. Artificial selection is intentional and has an end goal—the production of a dog breed with specific characteristics useful for a specific activity (hunting, sheep herding, protecting, and so forth). Natural

---

52. Eden, "Inadequacies of Neo-Darwinian Evolution," 11.
53. Bell, *Selection*, 199.

selection has no end goal. Can it do what artificial selection can do? Perhaps, but Bell has not demonstrated that it can.

Since Bell's work is theoretical rather than empirical, he appeals to John Endler's earlier work in order to make the case that natural selection is firmly established in natural populations. Bell cites the hundreds of studies that Endler collated in order to demonstrate selection in different circumstances. According to Bell, Endler's work "has brought to an end the period in which the ubiquity of natural selection could legitimately be questioned."[54] But as we have seen, in nearly half the studies, Endler rated knowledge of the selective agent as either questionable or unknown. Just how convincing are these studies? Let's consider a couple examples.

In 1984, Michael H. Fawcett published a study on the effects of predation on an herbivorous marine snail. At the southern end of the study area, there were two species of octopus that preyed on the snails, but neither octopus species existed in the northern zone of the study area. Fawcett found that the southern snails chose to live at a higher shore level, presumably to escape octopus predation, while the northern snails lived at a deeper level. Octopus predation was the likely selective agent that produced a change in the behavior of the southern snails.[55] Here, selection might account for a small change in snail behavior, but it could simply be that southern snails are aware of the existence of dangerous predators and try to avoid them, not that they have become genetically programmed to live higher in the shoreline because those that live deeper get eaten while those living higher are more likely to pass on their "higher-living" genes to their progeny. This is not a terribly convincing demonstration of natural selection in the wild.

In 1979, Fred Gould published a study based on laboratory research showing that a species of spider mite can evolve quickly in order to accept a new host plant that would have previously been toxic to the mite.[56] The experiment ran for twenty-one months, producing approximately fifty generations of mites. But it is difficult to understand how natural selection working on random variation would allow such a host switch (from lima bean to cucumber) since the earliest mites to colonize the toxic cucumber would all die since they had not yet evolved the mutation allowing them to survive on cucumber. Could some mites have been able

---

54. Ibid., 71.
55. Fawcett, "Local and Latitudinal Variation."
56. Gould, "Rapid Host Range Evolution."

to intentionally alter their genome to allow survival on cucumber? We will deal with this possibility in more detail in chapter 6 in a discussion of the concept of directed mutation.

A third example is not part of Endler's list but illustrates the difficulty of establishing the action of natural selection in wild populations. In 1996, Martin L. Cody and Jacob McC. Overton published a study involving seed dispersal distances in twenty-five species of plants that colonized islands in Barkly Sound from the nearby mainland around Vancouver, BC.[57] When plants colonize a small island habitat, a seed-dispersal system that disperses seeds a great distance is less advantageous than it is on the mainland, since on a small island a plant will likely be located close to water. Long seed-dispersal systems will guarantee that many seeds end up in the ocean rather than on land where they can germinate. Consequently, when plants colonize islands, they undergo rapid selection for reduced seed dispersal. Plants were studied on approximately two hundred small islands, and plant colonists from the mainland repeatedly developed reduced dispersal quickly after colonization. But if the variation on which natural selection acts occurs randomly, how is it that these plant colonists *repeatedly* received the proper mutation, which allowed them to propagate themselves successfully in an island habitat? Is this really a case of natural selection in action, or is there intentionality in the way the right mutation occurs at the right time to ensure survival? Cody and McC. Overton admit that "While a number of alternative explanations exist for the observed morphological shifts, all appear less likely than short-term evolution."[58] If there are alternative explanations, natural selection cannot be said to have been definitively demonstrated here.

These types of ambiguities plague many studies on natural selection in the wild and are a source of embarrassment for supporters of the grand narrative. Advocates of Darwinism are fond of responding to critics by emphasizing the mass of studies on natural selection filling the peer-reviewed journal literature. Surely the overwhelming volume of such studies is evidence itself of the reality of natural selection. Why would so many biologists be studying something for whose existence there is no evidence? For example, in a book written by scientists to counter theories of creationism and ID, Norman A. Johnson plays this very card:

---

57. Cody and Overton, "Short-Term Evolution."
58. Ibid., 60.

As a testimony to the extent of activity measuring evolution in the wild, Kingsolver et al. (2001) analyzed 2,500 estimates of selection gradients published in the technical literature between 1984 and 1997. Selection gradients are statistical estimators of the intensity of selection acting on a particular trait. They state that these 2,500 estimates are but a subset of the known studies chosen that satisfied rigorous criteria for inclusion in the study.[59]

Johnson simply emphasizes the sheer number of studies analyzed by Kingsolver and his collaborators as if the number of studies is by itself evidence for natural selection. But Johnson ignores the important caveats and conclusions discussed in the metastudy he cites.

Although Kingsolver and his collaborators did survey the results of 2,500 studies purporting to quantify the strength of selection in the wild, they caution readers that bias might undermine the data. First, natural populations are not chosen at random by biologists to study. Biologists only study populations where they already have a suspicion that natural selection might be at work. Second, since Kingsolver et al. only consider studies published in core peer-reviewed journals, which tend to exclude studies that fail to detect selection, they forthrightly state: "Indeed, there is clear evidence of some publication bias in our data set, in which estimates of weak and non-significant selection in studies with small sample sizes are relatively underrepresented. Each of these biases will tend to inflate our estimation of the strength of selection for some "random" trait or study system."[60] It is important to note the authors' own recognition that their results might inflate the strength of selection, because when they do draw conclusions about what the 2,500 studies show, we are told:

> Another important feature of the studies taken collectively is their low statistical power. Most of the studies were not replicated in either time or space. In addition, the sample size associated with estimates is quite low . . . implying that studies with smaller sample sizes tended to report relatively more estimates of selection. As a result, standard errors associated with the estimates are quite large relative to the magnitude of selection, and most studies lacked the statistical power to detect selection of "average" magnitude.[61]

---

59. Johnson, "Is Evolution 'Only a Theory'?" 356.
60. Kingsolver et al., "Strength of Phenotypic Selection," 253.
61. Ibid.

In other words, despite the number of studies being done to detect the action of natural selection in the wild, the quality of the studies and their ability to quantify selection are ambiguous at best, and even these modest conclusions may be inflated by bias. The authors finally conclude, "It is sobering that, for sample sizes exceeding approximately 1000, most estimates of linear and quadratic selection gradients cluster between -0.1 and 0.1: our most powerful studies indicate that selection is weak or absent."[62] In citing the Kingsolver article, Johnson has conveniently left out the important caveat and the mostly negative conclusion. The empirical evidence for natural selection in the wild is ambiguous at best.[63] But the grand narrative of Darwinian triumph consistently overrides this truth and presents natural selection as settled science. There is probably no better way to illustrate the ideological nature of the grand narrative and its effect on how we view natural selection than with a case study on the most iconic example of natural selection in the wild: industrial melanism in the peppered moth.

Industrial Melanism in the Peppered Moth

In England during the nineteenth century, naturalists became aware of an interesting phenomenon. Several species of moth were becoming darker in color, especially in and around major industrial centers like Manchester. To understand what might be causing this change, researchers began to focus on one species in particular—the peppered moth (*Biston betularia*). The typical form of the peppered moth has light-colored wings speckled with darker spots as if it had been sprinkled with pepper (hence the common name). A mutant form, known as *carbonaria*, was also observed; its wings were mostly dark in color. But until the middle of the nineteenth century, this darker form was comparatively rare. That changed quickly over the second half of the nineteenth century as the *carbonaria* form came to dominate the population, especially in industrial areas. This is the phenomenon known as *industrial melanism*.

Biologists quickly made the connection between the phenomenon of melanism in the peppered moth and the realities of British industrial

---

62. Ibid.

63. Similar kinds of ambiguous results have been obtained in other metastudies of the literature on natural selection in the wild. For example, see Hoekstra et al., "Strength and Tempo of Directional Selection in the Wild."

life. In the aftermath of the industrial revolution and the development of coal plants throughout England, it was clear that coal soot was beginning to pollute woodlands surrounding industrial centers. Tree trunks normally covered with lichens that gave the trunks a light color had become darker as industrial pollutants killed off the lichens. Biologists began to reason that the typical, lighter-colored moths would enjoy an advantage living against the lighter-colored, lichen-covered tree trunks since they would blend in and be invisible to predators, especially birds. But with the killing of the lichens and the darkening of the tree trunks by industrial pollution, darker-colored, melanic moths would have the survival advantage and would thus be expected to increase in frequency. In 1896, British researcher J. W. Tutt pronounced the phenomenon of industrial melanism to be an example of observable natural selection: according to Tutt, differential bird predation was the selective agent driving the evolutionary change. At a time when, as we have seen, Darwinian theory was in eclipse, having a clearly observable example of natural selection driving evolutionary change was a boon to Darwin's supporters, and not surprisingly, the bird predation theory of industrial melanism became the standard explanation.

Still, Tutt's idea was just a reasonable-sounding hypothesis; it lacked real empirical corroboration. Corroboration appeared to come in the 1950s when British physician and naturalist Bernard Kettlewell performed a number of now-famous experiments on the peppered moth. First, he set up cameras to record what happens when a population of peppered moths is released onto darkened, lichen-denuded tree trunks. Kettlewell recorded birds preying on the moths, noting that the birds seemed to have an easier time seeing the lighter-colored, typical variety of moth. He then performed a series of mark-release-recapture experiments. Kettlewell captured both typical, light-colored moths as well as dark-colored (*carbonaria*) moths, marked the undersides of their wings, and released them into a polluted woodland. He then set up traps to recapture as many as he could. In the process he recaptured about twice as many of the *carbonaria* forms as the typicals. He then repeated this experiment in an unpolluted woodland and received the opposite result, recapturing about twice as many typicals as *carbonaria*. It appeared that Kettlewell had confirmed Tutt's hypothesis with rigorous, controlled experiments demonstrating that industrial melanism was in fact an example of observable natural selection in action. Given the fact that melanic forms of the peppered moth began to decline after the implementation of antipollution legislation in the 1950s, the case appeared to be clinched. The peppered moth gained iconic status as the premier piece of evidence for Darwinian evolution, and the peppered moth case appeared in every biology textbook for the next half century.

But all was not well. As early as the 1960s, biologists began to notice major methodological flaws in Kettlewell's experiments. These are outlined fully in Michael Majerus's 1998 book *Melanism: Evolution in Action*. Most seriously, Kettlewell assumed that peppered moths rest during the day on tree trunks and therefore based his experiments on this assumption. But field observations showed this not to be the case. Peppered moths in the wild appear to spend their days resting on the underside of horizontal branches high up in the tree canopy where they are presumably better hidden from predators no matter their color. Kettlewell's experiments were not duplicating the experience of moths in the wild. He further assumed that bird vision is similar to human vision, that the degree of crypsis (how well hidden a moth appears against a certain background) would be the same for birds as for humans. But just because humans have trouble seeing a dark moth against a dark tree trunk does not mean birds would encounter the same difficulty. Research suggests that birds see differently than humans, and we cannot automatically evaluate the degree of crypsis using our own visual experience. Majerus concludes, "When the details of the genetics, behavior, and ecology of this moth are taken into account, the resulting story is one of greater complexity, and in many ways greater interest, than the simple story that is usually related."[64] Despite his frank and thorough discussion of the problems with Kettlewell's work, Majerus still registered belief in the bird predation theory as the most likely explanation for the phenomenon of industrial melanism. We will consider the basis for his belief below. For now, it is important to note that criticisms of Kettlewell's work were largely ignored before the publication of Majerus's book, and the peppered moth was therefore able to retain its iconic status.

But Majerus's book changed the equation. University of Chicago evolutionary biologist Jerry Coyne reviewed Majerus's book in 1998 for the journal *Nature*. He made several striking admissions. Coyne, a strong supporter of Darwinian theory, admitted to learning about the problems with Kettlewell's work for the first time while reading Majerus's book and confessed, "I unearthed additional problems when, embarrassed

---

64. Majerus, *Melanism*, 116.

at having taught the standard *Biston* story for years, I read Kettlewell's papers for the first time."[65] Finally realizing that the evidence for industrial melanism as a textbook example of natural selection in action was highly suspect, Coyne described his reaction as resembling "the dismay attending my discovery, at the age of six, that it was my father and not Santa who brought the presents on Christmas Eve."[66] Now recognizing the problematic nature of the evidence, Coyne concluded, "for the time being we must discard *Biston* as a well-understood example of natural selection in action, although it is clearly a case of evolution."[67]

Consider the significance of Coyne's confessions here. A PhD-trained biologist working at one of the premier research institutions in the world was unaware of the fundamental problems with a famous series of experiments because he had never read the original papers—yet the problems with Kettlewell's experiments had been known for nearly three decades! Coyne clearly realized how improbable this sounds and tried to rationalize it:

> It is . . . worth pondering why there has been general and unquestioned acceptance of Kettlewell's work. Perhaps such powerful stories discourage close scrutiny. Moreover, in evolutionary biology there is little payoff in repeating other people's experiments, and, unlike molecular biology, our field is not self-correcting because few studies depend on the accuracy of earlier ones. Finally, teachers such as myself often neglect original papers in favour of shorter textbook summaries which bleach the blemishes from complicated experiments.[68]

I leave it to readers to judge whether Coyne's rationalizations of his ignorance are convincing. But his last point deserves further comment because it provides a remarkable example of an idea discussed by Thomas Kuhn in *The Structure of Scientific Revolutions*.

Kuhn observed that science education differs from education in the humanities and social sciences. Education in the sciences rarely involves reading original works or pondering the significance of a science's historical development. Science students read textbooks instead, and textbooks function as initiations into the currently reigning paradigm under

---

65. Coyne, "Not Black and White," 35.
66. Ibid.
67. Ibid., 36.
68. Ibid.

which a particular branch of science operates. Kuhn says, "Textbooks thus begin by truncating the scientist's sense of his discipline's history and then proceed to supply a substitute for what they have eliminated."[69] Kuhn continues:

> For reasons that are both obvious and highly functional, science textbooks (and too many of the older histories of science) refer only to that part of the work of past scientists that can easily be viewed as contributions to the statement and solution of the text's paradigm problems. Partly by selection and partly by distortion, the scientists of earlier ages are implicitly represented as having worked upon the same set of fixed problems and in accordance with the same set of fixed canons that the most recent revolution in scientific theory and method has made seem scientific.[70]

If Kuhn is right, Coyne's reflections make more sense. With the success of the modern synthesis in giving Darwinism pride of place as *the* mechanism of evolution, and the compelling nature of the phenomenon of industrial melanism and Kettlewell's experiments as empirical verification of Darwinism, textbooks simply perpetuated the canonical story bleached of its blemishes. And Coyne simply fell victim to science's tendency to ignore original literature in the educational process. Kuhn's understanding of the workings of science certainly helps us make sense of Coyne's startling admissions, but we must not ignore the role ideology may have played here as well.

When Coyne's review of *Melanism: Evolution in Action* hit the public, antievolutionists had a field day. In an article appearing in the *Sunday Telegraph*, Robert Matthews wrote:

> Evolution experts are quietly admitting that one of their most cherished examples of Darwin's theory, the rise and fall of the peppered moth, is based on a series of scientific blunders. Experiments using the moth in the Fifties and long believed to prove the truth of natural selection are now thought to be worthless, having been designed to come up with the 'right' answer."[71]

---

69. Kuhn, *Structure of Scientific Revolutions*, 137.

70. Ibid., 138.

71. Matthews, "Scientists Pick Holes in Darwin Moth Theory," (cited in Majerus, "Peppered Moth," 375).

Creationists jumped on board and used Coyne's review as proof that Darwinism was dead. Coyne, a fierce opponent of creationists, had, by being honest, inadvertently emboldened his adversaries, and he clearly looked for an opportunity to redeem himself.

He did not have to wait long. In 2002, science writer Judith Hooper entered the moth wars with her book *Of Moths and Men: Intrigue, Tragedy, and the Peppered Moth*, giving Coyne an opportunity to jump right back into the fray with a review of this new book. While Hooper, like Majerus, raised critical questions about the peppered moth story, Coyne was far harsher in his review of Hooper than he had been in reviewing Majerus:

> The biggest shortcoming ... is Hooper's failure to emphasize that, despite arguments about the precise mechanism of selection, industrial melanism still represents a splendid example of evolution in action. The dramatic rise and fall of the frequency of melanism in *Biston betularia*, occurring in parallel on two continents, is a compelling case of evolution by natural selection. No force other than selection could have caused such striking and directional change.[72]

So in 1998, Coyne was ready to discard the peppered moth as a well-understood example of natural selection in action, but four years later trumpeted it as "a compelling case of evolution by natural selection." What accounts for this sudden reversal, especially considering no new evidence had come to light during those four years? Continuing his review of Hooper, Coyne writes: "This issue matters, at least in the United States, because creationists have promoted the problems with *Biston* as a refutation of evolution itself. Even my own brief critique of the story (*Nature* 396, 35–36, 1998) has become grist for the creationists' mill."[73] Coyne redeemed himself by pushing back against the creationists' use of his earlier review, but at the cost of demonstrating a striking inconsistency in his own views. What did Coyne really think about the peppered moth story?

In 2009, Coyne published *Why Evolution Is True*, a book that aimed to convince skeptical general readers that Darwinian evolution is both true and empirically confirmed. At one place, he states his desire for an example of natural selection in action: "we want to see a natural

---

72. Coyne, "Evolution under Pressure," 20.
73. Ibid.

population meet a natural challenge, we want to know what that challenge is, and we want to see the population evolve to meet it before our eyes."[74] Based on his 2002 statement we would expect Coyne to introduce the peppered moth story here. But interestingly, he never mentions the peppered moth, the phenomenon of industrial melanism, or Bernard Kettlewell anywhere in his book! Given the opportunity to double down on his 2002 affirmation of the importance of the peppered moth story, Coyne simply ignores it altogether. The view he expressed in 1998 clearly represents his true understanding—the peppered moth story needs for now to be discarded as an example of natural selection in action. His reversal in 2002 was surely motivated more by ideology than science. But this leaves us with a big question: Should industrial melanism in the peppered moth be discarded as a compelling example of natural selection in action? And if so, what does this do to the empirical status of natural selection as an evolutionary process?

In the same year that Majerus published *Melanism: Evolution in Action* (1998), Theodore Sargent, Craig Miller, and David Lambert published an article critically examining what they call the "classical" story about industrial melanism in the peppered moth. They arrived at many of the same conclusions as Majerus. But in addition, they were troubled by the speed with which the melanic moths took over the population, considering it potentially too fast to be explained by natural selection. Some early researchers had suggested the possibility that industrial melanism occurred not as a result of natural selection but rather as the result of induction—that either adult moths or moth larvae may have ingested industrial pollutants that turned their wings black. According to Sargent and his collaborators, "Such rapid changes would not seem so striking . . . if an induction process affecting many individuals, rather than a single mutation in a single individual, was the original source of the melanic moths."[75] They further observe that a number of studies show melanic moths dominating in unpolluted woodlands, noting that "these disparate cases pose problems for the 'classical' explanation of melanism."[76] Finally, they note that in the places where melanism is declining seemingly as a result of the passage of clean-air legislation, it is not clear that the lichens have grown back to significantly lighten the color of tree trunks.

74. Coyne, *Why Evolution Is True*, 132.
75. Sargent, "'Classical' Explanation of Industrial Melanism," 303.
76. Ibid., 304.

For these reasons and others, Sargent and his collaborators conclude that while the "classical" explanation might be true, "there is little persuasive evidence, in the form of rigorous and replicated observations and experiments, to support this explanation at the present time."[77] Yet recognizing the central role the peppered moth story has played in the history of evolutionary biology, they bow to the influence of the grand narrative anyway and say "we feel certain that this phenomenon is a product of selection."[78] On what is their certainty based, given that they have just spent twenty pages laying out multiple reasons to doubt such certainty? To try and save the "classical" story, they resort to speculation.

> What we are suggesting . . . is that the intuitive appeal of the "crypsis" explanation may have blinded us to the role that other selective factors *might* be playing in the melanism story. For example, strong selection *might* be operating on the life-cycle stages other than the adult. Thus, the larvae of typical and melanic morphs *might* differ with respect to such things as their tolerance of specific pollutants or their vulnerability to certain parasites or predators (who themselves *might* be responding to various effects of industrialization). In short, the complex of factors that *might* play a role in the increase (or decrease) of melanism in moths has barely been tapped.[79]

With repetition of the word "might," they try to suggest speculative scenarios that could save the peppered moth as an iconic example of natural selection in action. But they offer no evidence to support their speculative scenarios.

As I mentioned earlier, despite the thorough analysis of all the problems with Kettlewell's experiments that Majerus provides in his 1998 book, Majerus himself still wanted to view industrial melanism as an example of natural selection driven by differential bird predation. In a 2005 article, Majerus gives his primary reason for holding to this view. The passage is long but needs to be quoted in full.

> That said, my own conviction that bird predation is largely responsible is not based purely on empirical data from experiments published in the literature. I know that Tutt's differential bird predation hypothesis is correct because I know about peppered moths. For those who have never seen a peppered moth

---

77. Ibid., 318.
78. Ibid.
79. Ibid. (Italics added).

in the wild, which is almost everybody; for those with anti-Darwinian agendas; and for scientists, well-trained in rigour, stringency and experimental controls; for differing reasons this statement must seem insufficient, if not heretical. However, I stick by it.

The biography of the great geneticist Barbara McClintock was titled *A Feel for the Organism*. I think that I have a feel for certain organisms. My credentials are these: I caught my first butterfly when I was 4 years old. I learnt the basics of Mendelian genetics when I was 10 years old. For 45 years I have bred, collected, photographed and recorded moths, butterflies and ladybirds in the UK, across Europe and latterly around the world. I have run one or more moth traps almost nightly for 40 years. This experience has given me something of a feel for the organisms that I observe.

I bred my first broods of the peppered moth in 1964, following Ford's advice on careful separation of broods and writing notes on all procedures used. I found my first peppered moth at rest in the wild in the same year. As far as I am aware, I have found more peppered moths at rest in their natural resting position than any other person alive. I admit to being, in part, a moth man.

In the first chapter of *Of Moths and Men*, Hooper (2002) assassinates the character of 'moth men,' who have 'stunted social skills of the more monomaniacal computer hackers, going about with misbuttoned shirts and uncombed hair, spouting taxonomic Latin'. She cites Ted Sargent, who considers moth collectors to be weirder than butterfly collectors. According to Hooper, Sargent is awed by moth enthusiasts who 'can go up to a streetlight and start naming these things . . . it's an extraordinary talent'.

But it isn't extraordinary. Hundreds of thousands of children across the world can recognize hundreds of different Pokemon characters, and provide details of their characteristics, their evolutionary potential and their powers in contest. How is this different from a twelve-year-old who can recognize several hundred species of macro moth, know when they fly, and what their larvae feed upon? Calling out names to a group of people around a moth trap, the names I use are English, not Latin, for I learnt them, out of interest and fascination, when I was a child, and the English names were easier.

> I know the peppered moth, and I know that J. W. Tutt was essentially correct in his explanation of the rise of *carbonaria*.[80]

Unable to provide strong empirical support for the bird predation theory, Majerus falls back on an argument from authority. Because he knows the peppered moth, he knows industrial melanism is an example of natural selection. But scientific ideas are not proven by authority; they are proven by evidence. Majerus, like Sargent et al., desperately wanted the "classical" story to be true, but his support is clearly ideologically, not empirically, driven.

Majerus's is not the final word on industrial melanism in the peppered moth. In 2012, Geoffrey Fryer published "A New—and Retrospective—Look at Industrial Melanism in Moths" in the journal of the Linnean Society. According to Fryer, while everyone seized on J. W. Tutt's early support for the bird predation hypothesis to explain industrial melanism, they overlooked the observations made by another early British naturalist, observations that seemed to challenge the view that industrial melanism was an example of natural selection. In 1907, G. T. Porritt published a paper noting that several species of dark-colored moths moved from polluted woodlands into pristine woodlands where they thrived despite their inability to hide from predatory birds. Moreover, Porritt noted that some species of moth became paler rather than darker even in polluted woodlands, prompting Fryer to comment:

> The survival, and increased incidence, of pale moths in soot-blackened areas where, theoretically, they should have provided conspicuous targets for predatory birds (and been eliminated as, allegedly, were pale forms of those that became melanic) is an embarrassing anomaly for the orthodox story.[81]

Industrial melanism therefore might have nothing to do with natural selection but may instead result from some kind of disturbance to pigment synthesis or metabolism, possibly as a result of ingesting industrial pollutants.

This, as I previously mentioned, is known as the *induction theory* of industrial melanism. Ingestion of industrial pollutants, either at the adult or larval stage, could induce a mutation in many moths over a short time period, and these induced mutations could be heritable. The induction theory is certainly consistent with the observation that industrial

80. Majerus, "The Peppered Moth," 387.
81. Fryer, "New—and a Retrospective—Look," 32.

melanism declined after the passing of clean-air legislation both in industrial centers and in the pristine woodlands where "the invaders had been boosted by dispersal of adults from the original source and by small larvae windborne on threads of silk."[82] Clearly, the rush to find a convincing case of observable natural selection in the wild led biologists to seize on Tutt's work whole ignoring Porritt's. But when all the evidence is taken together, the idea that industrial melanism constitutes an example of natural selection in action is ambiguous at best; there are good reasons why Coyne ignores it altogether in his book arguing why evolution is true.

But supporters of the classical story have not gone quietly. Laurence Cook took issue with Fryer's article, wondering how the induction theory could explain the rapid decline in melanism after the 1970s since the reduction in industrial pollutants would not affect the mutants already in existence. Further, Cook argues that experiments designed to document the action of natural selection are not as flawed as widely believed. "Since the time of Kettlewell there have been a number of consistent independent trials showing an advantage to melanics in industrial areas, where the prevailing melanic frequency was 80 per cent or more."[83] But we should note that Cook along with I. J. Saccheri, in an article published in *Heredity* in 2013, concluded that "The peppered moth remains the type example of rapid response to human-induced environmental change, driven by selective predation."[84] But Cook and Saccheri, though giving a brief citation to Porritt, do not engage with any of his observations, and Fryer's 2012 article is ignored.

This prompted Fryer to respond once more with a detailed critique of Cook and defense of the induction theory. Fryer writes:

> While not proven, the induction of melanism by a mutagen can explain not only its widespread occurrence in industrial regions, irrespective of the habits of the moths concerned, be it to sit fully exposed or to be completely hidden; its spread into clean areas, for which natural selection cannot be invoked, and should in fact operate to prevent; and its later decline as levels of pollution—and the mutagen—fell. "Orthodox" mutation and natural selection do not.[85]

82. Ibid., 34.
83. Cook, "Fryer's New Look at Industrial Melanism in Moths," 12.
84. Cook and Saccheri, "Peppered Moth," 210.
85. Fryer, "How Should the History of Industrial Melanism in Moths Be

The phenomenon of industrial melanism continues to be a site of contestation among biologists. The bird predation theory seems attractive to anyone who wants an example of natural selection in action, but the available evidence does not appear to unambiguously support it. Induction is a possible alternative, but it has not yet been decisively proven either. So the question remains open, and its status as an open question has been known at least since the 1960s when the first criticisms of Kettlewell's experiments became widely known. But you would never know this by listening to the purveyors of the grand narrative of Darwinian triumph.

In 1985, Douglas Futuyma pointed to hundreds of studies showing how natural selection operates in the wild, listing the peppered moth as one example.[86] John Endler's 1986 book analyzed earlier in this chapter lists Kettlewell's experiments in the category of studies with a known selective agent. And in 1988, Ernst Mayr wrote, "The effect of selection would be far more convincing if one could demonstrate a selective value of specific components of the phenotype. And in fact this has been demonstrated by Kettlewell and others for the cryptic color of certain moths."[87] All of these references, and many more like them, ignore the ambiguities of the peppered moth story. Despite the evident difficulties, the peppered moth was perpetuated as the iconic example of natural selection in action and continued to be used in biology textbooks right up into the twenty-first century. The peppered moth may no longer be an icon of evolution, but it is a strong candidate to take its rightful place as the premier icon of evolutionary theory's grand narrative of Darwinian triumph.

Despite constant claims to the contrary from defenders of Darwinism, the modern synthesis largely failed to establish natural selection as the primary mechanism of evolution. Theodosius Dobzhansky attempted to draw a connection between the theoretical models of the population geneticists and studies of organisms in the wild, but he freely admitted

Interpreted?" 16.

86. Futuyma, "Evolution as Fact and Theory."

87. Mayr, *Toward a New Philosophy of Biology*, 137. See also Mayr's 1976 statement that "Industrial melanism (Kettlewell 1959) is a specially well-analyzed example of the fitness-enhancing property of single genes. Wherever soot darkens the bark of trees, the melanic moths gain a cryptic advantage over the normally pale-colored individuals of the species" (*Evolution and Diversity*, 94).

that confidence on the question of evolutionary mechanisms was not warranted in his day. George Gaylord Simpson attempted to draw a connection between the theoretical models of the population geneticists and the fossil record, but at best he only showed that the two *could* be understood as consistent, not that they *had* to be so understood. Later work by John Endler and Graham Bell only heightened the ambiguities involved in trying to empirically demonstrate the reality of natural selection in the wild while the iconic example of natural selection in action—industrial melanism in the peppered moth—remains uncertain even today.

Perhaps this should not surprise us. Even as staunch a defender of Darwinian evolution as Ernst Mayr would write in 1959:

> in spite of the almost universal acceptance of the "synthetic" theory of evolution, we are still far from fully understanding almost any of the more specific problems of evolution. I have tried to demonstrate this for every level from the chromosome up to the species and evolutionary line. There is still a vast and wide open frontier.[88]

That natural selection functioned as the primary mechanism of evolution was a proposition—an attractive one to be sure—but still just a proposition in the absence of a full understanding of the more specific problems of evolution. One of those specific problems concerns natural selection's ability—assuming natural selection occurs—to account for anything beyond small-scale changes occurring within populations of organisms, such as changes in coloration, antibiotic resistance in bacteria, or pesticide resistance in insects. Even if we accept that natural selection occurs in the wild and produces these microevolutionary changes, are there good reasons to think that natural selection can account for the kinds of macroevolutionary changes that led to the evolution of whole new classes of organisms? To this question we turn in the next chapter.

---

88. Mayr, "Where Are We?" 13.

5

Reframing the Modern Synthesis: Macroevolution

NATURAL SELECTION PROVED TO be a compelling idea in the abstract, but difficult to prove in its specifics. Despite the contrary claims made by the supporters of the grand narrative of Darwinian triumph, the modern synthesis largely failed to demonstrate the reality of natural selection as an important evolutionary process. Yet even if one grants the reality of natural selection and its ability to account for small-scale evolutionary change, natural selection can only be understood as "the mechanism of evolution" (as Graham Bell called it) if it can also account for large-scale evolutionary change (or macroevolution). Natural selection might be able to help bacteria develop resistance to an antibiotic or change the frequency of color morphs in moths (though even the latter is uncertain), but can it explain the evolution of whole new families, orders, and classes of organisms?

This question, according to the grand narrative, was definitively answered during the period of the modern synthesis by the German zoologist Bernhard Rensch in his canonical work *Evolution above the Species Level*. I will thus provide a critical engagement with Rensch's work followed by an analysis of Richard Goldschmidt's *The Material Basis of Evolution*, the most important challenge to Darwinian evolution written during the period of the modern synthesis. This will lead to a consideration of the Cambrian explosion (the sudden appearance of animal body plans during the Cambrian era) and polyploidy (the sudden appearance of new species by spontaneous genome duplication), two phenomena that challenge the grand narrative's emphasis on natural selection as the primary mechanism of evolution. How convincing is the connection between natural selection and macroevolutionary change?

Bernhard Rensch

Born in 1900, the German zoologist Bernhard Rensch (d. 1990) published *Neuere Probleme der Abstammungslehre* in 1947, a book credited with decisively demonstrating the power of natural selection to account for the evolution of higher taxonomic categories above the level of species. Rensch's book was included in the growing canon of the modern evolutionary synthesis and was later translated into English under the title *Evolution above the Species Level* (1960). I will analyze the English version in what follows.

The first thing to observe about Rensch's monograph is its unexpected polemical tone. Rensch is not satisfied to simply provide evidence demonstrating the sufficiency of natural selection to account for the evolution of higher taxonomic categories. In addition to his positive arguments, he repeatedly reminds his readers that there is no need to posit "autonomous forces" acting within organisms to explain the evolution of complex features. He repeats this polemical rejoinder so often that it becomes clear there is a larger ideological agenda at work. If the case for natural selection were convincing based on evidence alone, there would be no need to constantly polemicize against those who argue for "autonomous forces." Rensch's preference for natural selection seems to have been motivated in part by a desire to put a nail in the coffin of rival theories advanced by some of his fellow German scholars, including the geneticist Richard Goldschmidt (considered in more detail below) and the paleontologist Otto Schindewolf. Both held to the idea that macroevolution could not be explained by natural selection alone but (at least in the case of Schindewolf) by some kind of vital or autonomous force existing within organisms driving their evolution in specific directions. Rensch calls the idea of autonomous forces, "no explanation at all, as such assumptions would be only words denoting a process or a factor which cannot be analyzed or circumscribed physiologically."[1] Clearly, Rensch worried that positing autonomous forces might undermine the status of evolutionary biology as a naturalistic science. Given the constant

1. Rensch, *Evolution above the Species Level*, 98.

drumbeat of antivitalistic rhetoric characterizing his monograph, then, we must understand Rensch's advocacy of natural selection as occurring in a polemical environment that likely influenced his interpretation of the data. How convincing is his case that natural selection is a sufficient explanation for the evolution of organisms representing higher taxonomic categories?

Rensch begins by surveying research on the formation of geographical races within species (that is, he surveys within a species, variations dependent on geography) as well as the formation of species themselves. He is satisfied that "mutation, gene recombination, changes in population size, processes of selection and isolation, and occasional secondary hybridization is a sufficient explanation of all known types of races and species formation."[2] But on what grounds does Rensch make this claim? He surprisingly notes: "The special types of mutation effective in race formation have been analyzed in only a few cases, and in most animal forms it is hardly possible to analyze them, as many animal species are difficult to breed in captivity or their generations are too long to permit experiments."[3] Additionally, Rensch states, "All favorable mutants, especially those causing an increased viability or fertility, will increase in frequency as time passes, and hence races and species as a whole will change with time." But he then follows with, "In contrast to possible expectation, the number of paleontologic series proving such a historical race formation and speciation is not very large."[4] Rensch's confidence in natural selection to effect microevolutionary change seems betrayed by his frank admission of the lack of compelling paleontological evidence demonstrating this. He is right. The fossil record, as paleontologists well know, does not in many cases cooperate with the expectations set up by Darwin's theory.

Nevertheless, having "established" the sufficiency of natural selection to explain race and species formation, Rensch asks the central question of his book:

> Now the problem arises, whether these processes and the same factors can be considered a sufficient explanation of transspecific evolution leading to the emergence of new genera, families, orders, classes, etc., and hence to the formation of new organs

---

2. Ibid., 14.
3. Ibid., 16.
4. Ibid., 17.

and of new types of organization, or whether additional or even totally different factors must be assumed.[5]

Rensch, of course, will side with Darwin on this, but his willingness to force a Darwinian framework on the paleontological evidence will lead to some stunning inconsistencies in the balance of his monograph.

To understand this, some basic background in taxonomy is necessary. Biologists since the time of the Swedish zoologist Carl Linnaeus in the eighteenth century have organized living organisms according to a taxonomic hierarchy. Species stand at the bottom of the hierarchy, with similar species being grouped into genera; similar genera are grouped into families, families into orders, orders into classes, and classes into phyla. For example, human beings (*homo sapiens*) are classed as part of the chordate phylum (animals with a spinal cord), the mammalian class, the primate order, the hominid family, and the genus *Homo*. Species are the only classifications that actually exist in nature. The other categories are abstractions invented by biologists to organize the dizzying diversity of the living world.

Darwin's view of gradual, incremental change predicts that at the earliest stages of evolution the world would have been populated by a variety of closely related species. If there was a common ancestor, the species that diverged from that ancestor would have been closely related to that ancestor since evolution occurs gradually rather than by larger jumps (saltations). Only after long spans of time would species diverge enough that biologists would be able to sort them into different genera. And only after longer spans of time would the divergence increase enough for groups recognized as different families, orders, classes, and phyla to develop. In other words, Darwin's theory predicts that higher taxonomic categories could only come into existence after—and as a result of—a long period of divergence at lower taxonomic levels. Higher taxonomic categories would thus be younger than lower ones.

This, unfortunately for Darwin, is not what the fossil record indicates. At least among animals, higher taxonomic categories appear to have developed early with lower levels arising later to fill in the gaps between them. Higher categories thus appear to be older. Later in this chapter, we will consider this evidence in more detail in a discussion of the Cambrian explosion. Rensch, for his part, seems to recognize that higher taxonomic categories appear older than lower ones, but he ignores

---

5. Ibid., 57.

the non-Darwinian character of this evidence. In summarizing the evidence for the average age of higher taxonomic categories, Rensch writes:

> Generally, classes are twice as old as orders, orders are twice the age of families, and families are usually 1.5 times as old as genera. (Of course, there are many exceptions to these generalizations.) Genera are usually about ten times as old as species, and we have found that the average age of the species is somewhere between 100,000 and a few million years.[6]

This is an accurate characterization of the what the fossil record indicates, yet later, Rensch contradicts himself by observing that the fossil record shows evidence of rapid radiations of new species forming quickly, leading to the development of new higher taxonomic categories:

> As new families cannot be formed until new genera have differentiated, nor new orders until new families have originated, explosive phases of radiation lead to an increase not only of diverse forms but also of new structural types . . . So, for instance, during the Paleocene and Eocene there originated not only numerous mammalian genera, but also 102 new families and 19 new orders of animals.[7]

Rensch recognizes that on the basis of Darwinism, "new families cannot be formed until new genera have differentiated, nor new orders until new families have originated," meaning that the nineteen new orders of animals originating in this explosive radiation must be younger than the 102 new families, which in turn must be younger than the new genera. But this contradicts his previous statement that orders are twice the age of families and families 1.5 times as old as genera.

A few pages later, Rensch reverses himself again. Considering the transition of animal groups from life in the sea to life on land, he comments: "The change from aquatic to terrestrial life, requiring so many new adaptations, led to a simultaneous and rapid transformation in many animal groups. It is for this reason that in the Carboniferous 12 different orders of insects suddenly appeared."[8] But Darwin's theory does not allow for the "sudden" appearance of new orders. New orders arise only after long periods of divergence among species, giving rise gradually to new genera, new genera to new families, and new families to new orders.

6. Ibid., 91.
7. Ibid., 101.
8. Ibid., 109.

Darwin knew the fossil record contradicted expectations set up by his theory. Rensch seems to tacitly acknowledge this but ignores its larger significance, leading to major conceptual inconsistencies. Either higher taxonomic categories are older than lower ones, or they are younger. Rensch cannot have it both ways.

Given the rapid transformation of animal groups documented in the fossil record, do we know what drove this process? According to Rensch, "There are as yet no reliable data concerning the factors that caused the extremely rapid transformation of animal groups during the Early Silurian."[9] Without data, Rensch surmises that these rapid transformations are brought about by "a temporary intensification of selection due to environmental changes, e.g., by new types of vegetation or food resources, or due to the colonization of new ranges with habitats unoccupied or inhabited by types inferior in competition."[10] By focusing so much on the role of environmental factors driving rapid evolutionary transformations, Rensch sounds less like Darwin and more like Lamarck! But recall how William Bateson could find no correlation between environmental discontinuities and species distinctions. Rensch seems to have fallen into the trap articulated by Simpson: he has assumed the truth of natural selection and tried to force the data to fit it. But in many cases, it doesn't.

Rensch addresses another problem noted earlier by Bateson and others: the necessity of multiple coordinated mutations for maintaining the function of an organism during the evolutionary process. He argues that natural selection does not change an organism one trait at a time, but that the process of mutation creates systemic effects. Because of the factors of pleiotropy (a single gene can affect more than one trait) and gene interaction, Rensch argues that the process of mutation and selection can create the kinds of systemic effects that would allow an organism to maintain functional integrity as it evolved. Perhaps. But one's confidence in Rensch's explanation wanes when we consider the two examples of pleiotropy he provides. The first concerns a single gene mutation in the fruit fly that causes five alterations of the gross morphology of the fly. We might expect that these five alterations would render the fruit fly more fit in its environment, but instead we are told that "this gene ... is a

9. Ibid., 110.
10. Ibid., 112.

homozygous lethal, and hence has no evolutionary effect."[11] His second example concerns a mutation in the domestic hen that causes the feathers to curl, leading to a loss of body heat. From this, many other changes occur:

> The heart rate is increased and the ventricles are hypertrophic; the total amount of blood is increased; the spleen, pancreas, and kidneys are enlarged, and the intestine and cecum are lengthened. Besides these alterations, the white blood corpuscles are altered, the rate of metabolism is increased, the function of the thyroid gland is disturbed, there is a delay in ovulation and the structure of the testes is abnormal, from which decreased fertility and viability result.[12]

Rensch has succeeded in showing how single gene mutations can have systemic effects on an organism. But in both cases, the function of the organism is *degraded*, not enhanced, by the systemic effects, so these kinds of systemic changes would not be preserved by natural selection! If Rensch is unable to provide an example of pleiotropy leading to systemic effects that *enhance* an organism's viability, then he really cannot appeal to pleiotropy to solve the problem of maintaining functional integrity in evolution. Nevertheless, Rensch confidently asserts, "We may state that mutation and selection provide a sound basis for understanding even 'systemic" changes of whole organisms."[13] But assertion is not demonstration, and given the lack of positive evidence, one must remain skeptical.

Just as Rensch stands behind the widely held assumption (sans evidence) that pleiotropy maintains organismal integrity during evolution, so he also backs what's known as Cope's Rule—that most animal lineages increase in size as their populations evolve. This raises the question of directionality in evolution. Is there some built-in orthogenetic principle that causes organisms to become larger over time? On a Darwinian understanding, why would mutations creating a larger organism always be favored in every environmental circumstance? Wouldn't there be some circumstances where a smaller size might prove more beneficial? Rensch notes that indeed smaller size *might* be a favorable trait in some environmental contexts which is why Cope's Rule seems not to apply to the bulk of insect lineages. But for other types of animals, the assumption that larger size is virtually always favorable provides "a possible, though

11. Ibid., 130.
12. Ibid., 133.
13. Ibid., 167.

perhaps not fully sufficient, explanation of the phenomenon."[14] Any sense of directionality in evolution is anathema to strict Darwinian orthodoxy, and Rensch clearly recognizes the uncomfortable fit between this orthodoxy and the facts of the fossil record. But he nevertheless remains undeterred, concluding, "there is no need to suppose that Cope's Rule proves the existence of unknown autonomous forces of evolution; random mutation and natural selection provide sufficient interpretation of this kind of orthogenesis."[15] But the confidence of Rensch's pronouncement is betrayed by his statement that Cope's Rule provides only a possible but not an entirely sufficient explanation of this particular orthogenetic trend. Rensch simply assumes the sufficiency of natural selection as an explanation and ignores the fact that Cope's Rule does seem to raise questions about a Darwinian explanation of evolution. Even J. B. S. Haldane had earlier acknowledged the problematic nature for Darwinism of clear orthogenetic trends (like the tendency toward increasing size) documented in the fossil record.

This is not the only place where Rensch falls back on an argument by assumption. We see it again when he considers the standard explanation for the evolution of the complex interplay of the three bones that make up the inner ear of mammals—the malleus, incus, and stapes. These small bones are arranged in just such a way as to work perfectly in tandem to transmit vibrations and make hearing possible. It is hard to imagine how these three bones could have developed and come together in just this way through some gradual process of evolution where it would be necessary for each step in the process to confer a selective advantage on the organism. The mammalian ear only functions when all the bones are joined together in just the right way. How could earlier steps in a gradual process have been preserved by selection? Rensch echoes the standard evolutionary explanation that the three bones of the mammalian inner ear evolved from the three bones that make up the reptilian jaw. How does he know this? "The evolutionary course of all these accessory organs of the future mammalian ear, then, was by no means directed right from the first steps of transformation, and it may well be *interpreted by the assumption* that several mutational steps occurred, some of which were favored by selection."[16] Rensch's phrase "interpreted by the assumption"

---

14. Ibid., 218.
15. Ibid.
16. Ibid., 275 (italics added).

speaks volumes. Fossils gain meaning only when they are interpreted, and interpretation is always a subjective enterprise. If one starts from a Darwinian premise (as Rensch does), then of course, one can *imagine* a scenario by which several mutational steps transformed a reptilian jawbone into the mammalian inner ear. But imaginary scenarios prove nothing.

As with the bones of the inner ear, so with the eye, Rensch ignores the significance of evolutionary convergence as we discussed it in chapter 2, and boldly asserts:

> The fact that structures so different as the various types of eyes evolved independently and convergently, each representing a 'special structure' on the common basis of the transformation of epidermal nerve cells into photoreceptors, suggests that the evolution of organs was not brought about by autonomous forces, but by the directing influences of selection.[17]

Perhaps Rensch was unaware that St. George Mivart had made the reality of convergence one of his chief criticisms of Darwin's theory, or that Santiago Ramón y Cajal had confessed that the complexity of the eye had challenged his faith in Darwinism. Rensch is certainly within his rights to view convergence in eye evolution as supporting rather than challenging Darwin's theory. But his view would be more convincing if he had not at the same time said the following: "In all these examples, we cannot state with certainty in which stages of development the various steps of transformation occurred, but in numerous cases the *assumption* seems justified that the transformation took place by successive alterations of final morphogenetic stages."[18] Rensch places a heavy burden on assumption, but assuming how things may have happened is a far cry from demonstrating how they did happen.

It is fair to say that *Evolution above the Species Level* did not mount a convincing case for the sufficiency of natural selection to account for the evolution of higher taxonomic categories. By explicitly framing his work as a polemic against those who would propose "autonomous forces" at work in evolution, Rensch undermines his own scientific objectivity, lessening the credibility of his positive argument in favor of natural selection. He started from the assumption of natural selection's sufficiency (because what other *scientific* option is there?) and then tried to show how one

17. Ibid., 277.
18. Ibid. (Italics added).

could imagine natural selection leading to the evolution of new orders, classes, and phyla. But he certainly did not demonstrate that this is how these higher categories *did* develop. And unfortunately for both Rensch and Darwin, the fossil record indicates otherwise, as one of Rensch's chief polemical opponents knew so well. Richard Goldschmidt refused to fall in line with the emerging Darwinian orthodoxy of the synthesis. Are his reasons to demur any more convincing than Rensch's reasons for holding to the orthodox position?

Richard Goldschmidt

Born in 1878, Richard Goldschmidt (d. 1958) became the director of genetics at the Kaiser Wilhelm Institute for Biology in Berlin in 1913. In 1935, he fled the rise of Nazism and came to America where he taught at the University of California, Berkeley until 1948. Stephen Jay Gould called him "one of the premier geneticists of our century, a shoo-in on anybody's list of the top ten."[19] In 1940, Goldschmidt published *The Material Basis of Evolution*, a major work in which he formulated a dissenting argument on the question of whether macroevolution can be explained by the same processes leading to microsevolutionary change. Given Goldschmidt's preeminent status within the international community of biologists, his challenge to the sufficiency of Darwinism is noteworthy. It constitutes a real problem for the grand narrative. It is simply not the case that all biologists rallied around Darwinism during the period of the modern synthesis. Goldschmidt refused to play along, much to the chagrin of some of his contemporaries like Ernst Mayr, as we will see.

Similar to Thomas Hunt Morgan, Goldschmidt could not see how the normal range of hereditary variation found in populations of organisms could have any meaning for an understanding of large-scale evolution. These kinds of variations could be marshalled by natural selection to create diversification and adaptation within species, but "it is always the same little change of tune which is produced in these cases, melanisms, albinisms, rutilisms, and their like." These minor changes, Goldschmidt thought, were important "only as material for genetic experimentation."[20] They were too insignificant to be the basis for large-scale evolutionary change.

19. Gould, "Uses of Heresy," xv.
20. Goldschmidt, *Material Basis of Evolution*, 27.

Of course, microevolutionary changes stood at the heart of the work of the population geneticists, and as a geneticist himself, one might expect Goldschmidt to have had a high regard for the work of his population geneticist colleagues. This appears to have been the case, for Goldschmidt in one place described Fisher, Haldane, and Wright's work as "brilliant." Yet he also felt it important to note that "biology must be studied *with* mathematics but not *as* mathematics. This means that the most brilliant mathematical treatment is in vain if the biological rating of the material is not correct."[21] He continues:

> I am of the opinion that this criticism applies also to the mathematical study of evolution. This study takes it for granted that evolution proceeds by slow accumulation of micromutations through selection, and that the rate of mutation of evolutionary importance is comparable to that of laboratory mutations, which latter are certainly a motley mixture of different processes of dubious evolutionary significance. If, however, evolution does not proceed according to the neo-Darwinian scheme, the mathematical study turns out to be based on wrong premises.[22]

Like Rensch, Goldschmidt was well aware that the fossil record did not square with the expectation set up by Darwinian evolution that higher taxonomic categories should come into being later rather than earlier, and he took this as serious evidence against a Darwinian view, while Rensch essentially ignored it. For Goldschmidt, the available evidence did not support the Darwinian scheme, rendering the mathematical work of the population geneticists moot, even if the mathematical formulations themselves *were* brilliant. Macroevolution must occur as a result of another evolutionary process than just sheer accumulation of micromutations. But what kind of process?

Goldschmidt proposed that occasionally an organism would be born having undergone during its embryonic development a large-scale repatterning of its chromosomes, leading to a significant morphological change. He writes: "A systemic mutation (or series of such), then, consists of a change of interchromosomal pattern. This is what is actually found taxonomically (the bridgeless gap) and cytologically. Whatever genes or gene mutations might be, they do not enter this picture at all."[23] Because

---

21. Ibid., 137 (italics original).
22. Ibid.
23. Ibid., 206.

species and higher taxonomic categories seem to be characterized by bridgeless gaps (dogs and cats are clearly distinct kinds of organisms, for example), whatever evolutionary process created this kind of discontinuity in organisms had to be a discontinuous process itself. Since Darwinian evolution was a continuous process, it could not, almost by definition, produce discontinuous evolutionary change. But systemic mutations—a major repatterning of the chromosomes—might.

Given that Goldschmidt was a geneticist by training, it is noteworthy that he came to question the very existence of individual genes and developed a more dynamic view of the genome as a complex and integrated system. On this he might have been prescient. In a 2013 article, Stuart Newman pronounced the "Demise of the Gene" to point out how modern biology increasingly demonstrates the vast gap that stands between our knowledge of genetics and our knowledge of how organisms develop their specific phenotypes (i.e., observable traits). Unraveling the full genetic library of an organism (which has now been done for many organisms) simply does not tell us much about the form those organisms take. Predicting what an organism will look or act like based on an analysis of its genes simply cannot be done. For example, to emphasize our close evolutionary relationship to our ape ancestors (and to needle their creationst opponents) Darwinians regularly rehearse the fact that humans and chimpanzees share 98 percent of their DNA. But rarely do we stop and consider why two organisms 98 percent identical at a genetic level are still so vastly different at an organismal level. Positing evolution as merely the changes in gene frequencies over time (Darwinism), Newman argues, appears to be a fool's errand.[24] Given these modern understandings, Goldschmidt's challenge to Darwinism cannot be easily dismissed, though, as we will see, advocates of the grand narrative will certainly try to dismiss it.

In an early paper Goldschmidt coined the term "hopeful monster" to refer to those rare events when a systemic repatterning of the chromosomes would lead to the emergence of a fundamentally new kind of organism. "Hopeful monster," not surprisingly, became a term of derision in the hands of Goldschmidt's opponents, but he was always willing to defend it. He points out that for dog breeders, a dog with very short bowlegs would have been seen as a monstrous mutant until a proper niche for it was found—to follow a badger (*ein Dachs*) into its den, thus

---

24. Newman, "Demise of the Gene."

producing the Dachshund.[25] Goldschmidt reminds us that even Darwin pointed out how monstrosities sometimes occur under domestication, so the "hopeful monster" was certainly not a new idea. Still, he realized the biggest problem with his theory was a lack of evidence that such systemic mutations had occurred in such a way as to produce new and viable forms of life. But Goldschmidt had a ready answer: "It would be very cheap criticism, indeed, to say that nobody has ever witnessed such a process. Neither has anyone witnessed the production of a new specimen of a higher taxonomic category by selection of micromutations."[26] Goldschmidt realized his theory was speculative, but it was no less so than Darwinian orthodoxy. No one has ever witnessed the emergence of a new higher taxonomic category by any means. But you would never know this by reading only the canonical literature of the grand narrative. Over time, and despite the great respect with which Goldschmidt was held within the biological community in the first half of the twentieth century, he became a derided figure precisely because his challenge to Darwinian orthodoxy proved so difficult to dismiss on the grounds of evidence. How was he received during the period of the synthesis and in the postsynthesis years?

Just as he had done with Thomas Hunt Morgan, Ernst Mayr also labeled Goldschmidt an essentialist who did not understand population thinking. This of course was Mayr's standard criticism of anyone who rejected Darwinism, so we should allow ourselves appropriate skepticism about this criticism's legitimacy. Goldschmidt's refusal to assent to developing Darwinian orthodoxy was a real source of frustration for Mayr, both because Mayr knew Goldschmidt personally, and because Mayr had great respect for him as a scientist, calling him in one place "a distinguished biologist"[27] and in another "one among a group of outstanding authorities."[28] Mayr relates a day he spent with Goldschmidt at the New York Museum of Natural History in 1936 or 1937 when they discussed the topic of *geographic speciation*. Geographic speciation is the Darwinian idea that new species arise when a population of organisms is split into two separated subpopulations such that the two populations then evolve in isolation from each other until they diverge into separate

---

25. Goldschmidt, *Material Basis of Evolution*, 391.
26. Goldschmidt, "Evolution, as Viewed by One Geneticist," 97.
27. Mayr, *Animal Species and Evolution*, 481.
28. Ibid., 586.

species. The idea of geographical speciation stood at the center of Mayr's work, but try as he might he could never convince Goldschmidt to accept it. Mayr comments that until Goldschmidt delivered the Silliman Lectures at Yale University in 1939 (lectures that became the basis of *The Material Basis of Evolution*), "I had simply taken it for granted that anybody who seriously thought about the subject would have to adopt the concept of geographical speciation."[29] But upon learning that this was not the case, he continued: "Even though personally I got along very well with Goldschmidt, I was thoroughly furious at his book, and much of my first draft of *Systematics and the Origin of Species* was written in angry reaction to Goldschmidt's total neglect of such overwhelming and convincing evidence."[30] The evidence may have seemed overwhelming and convincing to Mayr, but clearly it did not seem so to Goldschmidt, much to Mayr's chagrin.

Theodosius Dobzhansky had a somewhat different view. He penned a review of *The Material Basis of Evolution* for the journal *Science* in which he viewed Goldschmidt's book as marking "an at least temporary end of the undivided reign of neo-Darwinian theories."[31] Dobzhansky goes on to commend Goldschmidt's splendid command of multiple biological disciplines, calling his theory both "brilliantly developed" and "masterfully presented." Though he thinks Goldschmidt's theory of systemic mutations is invalid, "it must be admitted that Goldschmidt has marshalled an impressive array of evidence in its favor."[32] Dobzhansky then concludes his review with this:

> But in the reviewer's opinion the simplicity of Goldschmidt's theory is that of a belief in miracles. It must, nevertheless, be recognized that Goldschmidt's keenly critical analysis has emphasized the weaknesses and deficiencies of the neo-Darwinian conception of evolution, which are numerous, as even partisans ought to have the courage to admit.[33]

As we saw in the previous chapter, Dobzhansky seems never to have missed an opportunity to emphasize the weaknesses and deficiencies in

---

29. Mayr, "How I Became a Darwinian," 420.
30. Ibid.," 421.
31. Dobzhansky, Review of *The Material Basis of Evolution*, 356.
32. Ibid., 357.
33. Ibid., 358.

Darwinism! Sewall Wright, who also reviewed Goldschmidt's book, essentially agreed with Dobzhansky:

> While the reviewer radically disagrees with the author's central thesis, he wishes to testify to the importance of the book. A great store of well-selected data have been assembled from diverse sources, fairly presented and discussed from viewpoints which must be carefully considered by anyone interested in the problem of evolution.[34]

As Goldschmidt challenged the emerging Darwinian orthodoxy of the modern synthesis, the shapers of the synthesis could not ignore his trenchant critiques. This contrasts dramatically with how Goldschmidt would come to be treated in postsynthesis years.

Consider Stephen Jay Gould's reflections on his experience as a college student studying biology in the 1960s:

> I had never heard of Richard Goldschmidt. Yet his name surfaced in almost every course—never with any explication of his views, but only in a fleeting and derisive reference to something called a "hopeful monster." Students then responded with a derisive sign of recognition—as our professors seemed to expect as a badge of membership in some inner circle. I found the oft-repeated exercise—one might almost call it a ritual—offensive and demeaning, both to Goldschmidt and to any notion of my potentially independent intelligence.[35]

Gould's experience was not unique, for he relays the similar reflections of T. H. Frazzetta:

> No one stopped to consider whether in all of Goldschmidt's assailable propositions, there existed anything worth thinking about. There was no time for such consideration as long as there was so much merry mayhem to be carried out. In my university classes, the name 'Goldschmidt' was always introduced as a kind of biological 'in joke,' and all we students laughed and snickered dutifully to prove that we were not guilty of either ignorance or heresy.[36]

And there was also Guy Bush: "When his name did come up it was inevitably in the context of 'hopeful monsters' and to the accompaniment

---

34. Wright, Review of *The Material Basis of Evolution*, 170.
35. Gould, *Structure of Evolutionary Theory*, 452.
36. Cited in ibid., 452.

of subdued snickers and knowing nods. It didn't take long to learn that Richard B. Goldschmidt was not to be taken seriously as an evolutionary biologist."[37] But Goldschmidt *was* taken seriously in the 1940s. However, when it became impossible to counter his incisive criticisms of Darwinian orthodoxy on the basis of evidence and argument, the tradition simply ridiculed him and tried to ignore him altogether. As Gould notes, "Every orthodoxy needs a whipping boy,"[38] and Goldschmidt was one for believers in the grand narrative of Darwinian triumph. But as any religion scholar knows so well, ideas are not rendered true by virtue of being rendered orthodox. Orthodoxies emerge because they serve political, economic, and social interests. Interestingly, despite the widespread denigration of Goldschmidt, not all later biologists treated his work in so patronizing a way.

In 1975, Hampton Carson concluded an article on the genetics of speciation with the words, "Speciational events may be set in motion and important genetic saltations toward species formation accomplished by a series of catastrophic, stochastic genetic events."[39] Carson is referring to an idea very similar to Goldschmidt's systemic mutations. In 1980, Viktor Hamburger hailed Goldschmidt for breaking new ground in supplementing traditional views of genetics with "a dynamic conception of the gene as a physiological agent controlling developmental processes. By creating this link, he opened a meaningful dialogue between some embryologists and some geneticists."[40] As the study of embryological development was largely left out of the modern synthesis, Goldschmidt's openness to developmental processes as evolutionarily significant may also be a reason for his marginalization by orthodox Darwinians. Finally in 1992, K. S. Thomson called Goldschmidt's book more interesting than its usual caricature, and suggested that the term "hopeful monster" was merely a way to state the problem with macroevolution: how to account for large-scale evolutionary change without appealing to the neglected issue of embryological development.[41] The Darwinian view, Thomson says,

37. Ibid.
38. Ibid.
39. Carson, "Genetics of Speciation," 87.
40. Hamburger, "Embryology," 105.
41. Thomson, "Macroevolution," 111.

requires us to posit "the origin of the magnificently improbable from the ineffably trivial."[42] Goldschmidt could not have said it any better.

Even if one does not accept Goldschmidt's concept of systemic mutations as providing the raw material for large-scale evolutionary change, his recognition of the problem with Darwinian evolution is right on point. Rensch may have tried to sidestep the issue, but the fact is that the fossil record does suggest that at least among animals, higher taxonomic categories did evolve earlier than lower taxonomic categories, in complete opposition to the expectations of Darwin's theory. This phenomenon is widely known as the Cambrian explosion, an extremely important event in the history of life, which seems much more consistent with Goldschmidt's saltational understanding of evolution than it does with Rensch's gradualist Darwinian view. Darwin knew the Cambrian explosion might prove fatal to his theory. He was right to be so concerned.

The Cambrian Explosion

When he wrote *On the Origin of Species*, Charles Darwin was well aware that fossils of very early animal forms found in Cambria, Wales, seemed to contradict his theory. In a section of the *Origin* subtitled "On the sudden appearance of Groups of allied Species in the lowest known Fossiliferous Strata," Darwin puzzled over how to explain what appeared to be a rather sudden emergence into the fossil record of a variety of animal forms representing higher taxonomic categories without a record of gradual diversification leading up to these representatives of higher taxa. On the absence of the expected pre-Cambrian fossils that would link to the new Cambrian forms he commented, "To the question why we do not find rich fossiliferous deposits belonging to these assumed earliest periods prior to the Cambrian system, I can give no satisfactory reason."[43] Darwin hoped that what appeared to be a near fatal flaw in his theory was only apparent and resulted from the imperfection of the fossil record. As we established in chapter 2, Darwin was a master at explaining away difficulties with his ideas, and no less so here:

> I look at the geological record as a history of the world imperfectly kept, and written in a changing dialect; of this history we possess the last volume alone, relating only to two or three

---

42. Ibid., 107.
43. Darwin, *On the Origin of Species*, 253.

countries. Of this volume, only here and there a few lines. Each word of the slowly-changing language, more or less different in the successive chapters, may represent the forms of life, which are entombed in our consecutive formations, and which falsely appear to have been abruptly introduced. On this view, the difficulties above discussed are greatly diminished, or even disappear.[44]

Darwin hoped that new fossil discoveries in the years ahead would vindicate him and show that organisms representing higher taxonomic categories had not in fact appeared rather suddenly in the Cambrian period but rather had emerged through the long, steady divergence of pre-Cambrian forms. But he was clear: "The case at present must remain inexplicable; and may be truly urged as a valid argument against the views here entertained."[45]

Had Darwin lived on, he would have been sorely disappointed. In 1909, Charles Walcott, director of the Smithsonian Institution, discovered an enormous collection of exquisitely preserved Cambrian-era fossils in the Canadian Rockies in a geological formation known as the Burgess Shale. Later in the twentieth century, the Burgess Shale fossils were supplemented by two additional large Cambrian fossil collections: the Chengjiang fauna from Yunnan, China, and the Sirius Passet fauna from Greenland. What these fossil collections show, contrary to the expectations set up by Darwin's theory, is that beginning about 530 million years ago, organisms representing more than two dozen distinct animal phyla evolved rapidly (within perhaps a span of ten million years) without having emerged through a long, steady, gradual process of low-level diversification leading up to this Cambrian explosion of new animal body plans. Just before the Cambrian period, Earth's oceans were inhabited by small, fairly simple soft-bodied organisms. But suddenly (geologically speaking) in the Cambrian era, the oceans teemed with a veritable menagerie of new, more complex creatures, including the first with skeletal structures. They are strange-looking creatures with equally strange names: *Opabinia, Anomalocarus, Wiwaxia,* and *Pikaia* (which might count as the first chordate, the phylum to which humans belong). These strange creatures no longer exist, but many of the animal phyla they represent do. Almost all contemporary animal phyla—vertebrates, invertebrates, mollusks, segmented worms, jellyfish, and many

44. Ibid., 255.
45. Ibid., 254.

others—seem to have developed rapidly in the Cambrian explosion with virtually no new animal phyla having evolved since. Darwin was right. If the Cambrian explosion was a real event, his theory of natural selection cannot by itself explain the evolutionary history of life on Earth. Let me be clear about why.

As we saw in the earlier discussion of Bernhard Rensch's work, Darwinian evolution proceeds gradually as closely related species diverge from each other driven by the selection of very small variations. Only after enormous spans of time will the evolution of new species diverge enough for what we recognize as higher taxonomic categories to come into existence. The divergence of species leads to the development of distinct genera, genera to the development of families, and so on up the taxonomic hierarchy to phyla. It is the evolution of lower-level taxonomic categories that leads to the evolution of higher taxa. So in no case can a Darwinian process produce phylum-level distinctions first, and then fill in each of these phyla over time with lower-level categories. But this is exactly what the Cambrian explosion seems to show happened 530 million years ago—but never before or since. Instead of species gradually diverging into new, closely related species, species somehow rapidly diverged into new organisms representing fundamentally different body architecture. Only then over the next 530 million years did the evolutionary process fill in these sparsely populated phyla with the great variety of classes, orders, families, genera, and species that characterize the animal world today. This is the opposite of what natural selection should produce and raises the question of whether the Cambrian explosion was a real biological event or simply an apparent one due to poor preservation of pre-Cambrian fossils.

The majority of biologists who have examined the evidence consider the Cambrian explosion to have been a real biological event. In a 1992 *Scientific American* article, "The Big Bang of Animal Evolution," Jeffrey Levinton attests to the mysterious nature of the explosion.[46] The extraordinary nature of this event was seconded in 1999 by Andrew H. Knoll and Sean B. Carroll in a study on early animal evolution published in *Science*: "Cambrian diversification . . . reflects an evolutionary milestone regardless of the length or character of earlier animal history."[47] More recently, Douglas H. Erwin and James W. Valentine, in their textbook

---

46. Levinton, "Big Bang of Animal Evolution."
47. Knoll and Carroll, "Early Animal Evolution," 2130.

The Cambrian Explosion: The Construction of Animal Biodiversity, write, "Several lines of evidence are consistent with the reality of the Cambrian explosion."[48] Even Ernst Mayr had little choice but to acknowledge the extraordinary nature of this event:

> This period of seemingly exuberant production of new structural types (phyla) soon came to an end. Altogether some 70 or 80 different structural types (body plans) appeared in the late Precambrian and Cambrian, but apparently no new ones originated at any later period . . . All currently living phyla of animals, about 35 of them, were long thought to have originated during a period of only about 10 million years in the early Cambrian. How could one explain such short-lived exuberance of structural innovation at that period?[49]

Not surprisingly, Mayr tries to "defuse" the explosion by, like Darwin, appealing to the likely poor preservation of fossils and by using molecular clocks to date divergence times of taxonomic categories. I will return to these two issues below, but first we need to consider a study that highlights the entirely unprecedented nature of the Cambrian explosion.

The Cambrian explosion was not the only large-scale diversification of animal life that has occurred during Earth's evolutionary history. About 250 million years ago, at the end of the Permian period, Earth underwent a major extinction spasm in which nearly 96 percent of all marine life disappeared with about 70 percent of terrestrial life. This was by far the largest mass extinction ever to occur—90 percent of all living things perished. Such a large-scale extinction event created an opportunity for life to diversify again in its aftermath. In 1987, Douglas Erwin, James Valentine, and John Sepkoski published a study comparing the diversification of animal life characterizing the Cambrian explosion with the diversification characterizing this post-Permian extinction. The results are noteworthy. Although a large number of new phyla emerged in the Cambrian period, no new phyla appeared after the Permian extinction. This latter diversification was characterized only by the emergence of new lower-level taxonomic categories. They write:

> The fossil record suggests that the major pulse of diversification of phyla occurs before that of classes, classes before that of orders, and orders before that of families. This is not to say

---

48. Erwin and Valentine, *Cambrian Explosion*, 6.
49. Mayr, *What Evolution Is*, 59.

that higher taxa originated before species (each phylum, class, or order contained at least one species, genus, family, etc. upon appearance), but the higher taxa do not seem to have diverged through an accumulation of lower taxa. Instead, the lower taxa appear to be exploiting the potentialities of the novel body plans recognized as higher taxa in the relatively empty adaptive space of the Early Cambrian.[50]

Despite the fact that the earth was nearly wiped clean of life after the Permian extinction, the reemergence of life continued to perpetuate the same phylum-level categories that had developed in the Cambrian period. With an almost unlimited ability for life to diversify once again, all this diversification occurred in lower taxonomic levels within already existing phyla. The rapid emergence of new phyla in the Cambrian explosion was a truly unique event in the history of life, and one for which Darwin's theory seems unable to account. But this has not stopped its proponents from trying.

How confident can we be that there was not a long period of low-level diversification in the pre-Cambrian period leading to the appearance of new phyla—a period of diversification that simply has not been preserved in the fossil record, making the Cambrian event appear to be an explosion when it was in fact a slow, gradual, Darwinian process? As we saw above, Mayr appeals to poor fossil preservation as a way to reduce the impact of the explosion. It is important to note that animals did not first appear during the Cambrian period. Fossil animals have been found in pre-Cambrian sediments (most notably in Australia) known as the Ediacaran fauna. But these are small, soft-bodied, and fairly simple organisms that do not display the complexity of the Cambrian fauna. A large gap still stands between the two, and it seems likely that since the soft-bodied organisms of the Ediacaran fauna fossilized, there would be a fossil record of organisms filling the gap between the Ediacaran and Cambrian fossils. Yet numerous biologists have concluded that the fossil record, while far from perfect, is complete enough to give an accurate portrait of life in the Ediacaran and Cambrian periods.[51] It appears the Cambrian explosion was a real biological event.

Another way that Darwinian proponents try to defuse the explosion is through an appeal to molecular clock dating. By analyzing DNA

---

50. Erwin et al., "Comparative Study of Diversification Events," 1183.

51. See Erwin and Valentine, *Cambrian Explosion*, 144; Knoll and Carroll, "Early Animal Evolution," 2130.

or protein sequences in living organisms it may be possible to determine when past organisms representing different taxonomic categories diverged. This assumes that DNA or amino acid substitutions occur at a predictable rate. If they do, and molecular analysis can be calibrated based on known divergence times via the fossil record, then molecular analysis may be able to provide divergence times for groups of organisms not well preserved in the fossil record. This is exactly what some biologists have done for the Cambrian explosion. Molecular clock analyses consistently place divergence times of animal groups much earlier than the fossil record documents. This is interpreted as evidence that a long, gradual period of diversification leading up to the Cambrian explosion did in fact occur even if the fossil record fails to preserve evidence of it.

But molecular clock dating has come under considerable scrutiny, and many question its accuracy. Molecular clocks often give widely divergent dates depending on what DNA or protein sequences are used. In the view of James Valentine, David Jablonski, and Douglas Erwin, "the accuracy of molecular clocks is problematical, at least for phylum divergences, for the estimates vary by some 800 million years depending on the technique and molecules used."[52] Francisco Ayala states, "We lack a valid theory that would allow us to calculate the probable error of estimates based on the molecular clock," and concludes, "The molecular clock is sloppy, but we do not yet know how sloppy it is."[53] Fossils are the only hard evidence we have of organisms living in the distant past. It is difficult to argue that hypothetical organisms inferred from molecular clock dating should trump what the fossil record actually documents about the history of life on Earth. If we follow the fossil record, the Cambrian explosion stands as a unique challenge to Darwinian evolution.

Erwin and Valentine point out other important aspects of the Cambrian explosion to emphasize its unique importance in the history of animal evolution. For example:

> Sponges invented many basic functional elements of metazoan development, expanding the genome from their unicellular inheritance. As this evolutionary accomplishment probably took close to 100 million years, it provides some perspective on the evolutionary abruptness of the rise of the disparate bilaterians

---

52. Valentine, et al., "Fossils, Molecules, and Embryos," 856.
53. Ayala, "On the Virtues and Pitfalls," 413.

that composed the bulk of the Cambrian explosion fauna, which evidently took just a few tens of millions of years.[54]

If it took 100 million years to evolve an organism as simple as a sponge, it is hard to imagine how the vastly greater complexity documented in explosion fauna evolved in only ten million to twenty million years. And given our discussion of eye evolution in chapter 2, it is significant to note evidence that "the recent discovery of exquisitely preserved eyes from arthropods in the early Cambrian Emu Bay Shale in Australia illustrates that highly advanced, compound eyes . . . had evolved very early in the history of the clade."[55] Even something as complex as the eye may have evolved quickly, rather than through a long, gradual, Darwinian process.

This raises the important question of what caused such a unique evolutionary event. Can the microevolutionary processes normally studied by biologists in existing organisms explain the Cambrian explosion, or do we need to appeal to additional macroevolutionary processes? Erwin and Valentine hold to the latter position.[56] But what might those additional processes be? They have little to offer here. Although they appeal to vague notions of the "construction of ecological interactions and the introduction of new methods of developmental regulation,"[57] this is really just a statement of the obvious. We know that new methods of developmental gene regulation would have to evolve to produce such a stunning flowering of complexity, but how did these new methods of regulation themselves evolve, and in such a short period of time? Without a clear answer to this question, Erwin and Valentine appropriately conclude:

> Clearly, the biosphere has promoted its own evolutionary trajectory, and the Cambrian explosion was a once-in-an-era happening; it could hardly have been more complicated and could hardly be more tantalizing. In addition, there can hardly be more of a challenge to paleobiologists, evolutionary biologists, and many other scientists than to describe and interpret the confluence of history and process responsible for events during that remote and critical time in life's history.[58]

54. Erwin and Valentine, *Cambrian Explosion*, 270.
55. Ibid., 216.
56. Ibid., 10.
57. Ibid., 337.
58. Ibid., 342.

Tantalizing it is. And this raises the question of why so few people know about the Cambrian explosion. The Cambrian explosion is to evolutionary biology what the Dead Sea Scrolls are to biblical studies. But while the average person has at least heard of the Dead Sea Scrolls and may have a passing knowledge of their importance, few people outside biological circles have ever heard the term *Cambrian explosion.* Stephen Jay Gould asks, "Why is *Opabinia*, a key animal in a new view of life, not a household name in domiciles that care about the riddles of existence?"[59] The answer, of course, is that when evolution is packaged for the general public, the significance of the Cambrian explosion is either explained away or ignored. The grand narrative of evolutionary triumph simply has no good way to handle this non-Darwinian, inconvenient truth.

We have already seen above how Ernst Mayr tried to explain away the Cambrian explosion by appealing to poor fossil preservation and molecular clock dating. In other places, he acknowledges the reality of the explosion but simply ignores its significance. In discussing the question of why new structural types have not evolved since the Cambrian, he argues that what he calls cohesion of the genotype makes it very difficult to generate the kinds of new, complex biological information necessary to create new kinds of body plans. Once the genotype supporting a particular body plan is set, it is difficult to break out of that straitjacket. According to Mayr:

> This is the reason why only so relatively few new structural types have arisen in the last 500 million years, and this may well be the reason why 99.999% of all evolutionary lines have become extinct. They did so because the cohesion prevented them from responding quickly to sudden new demands by the environment.[60]

This may be true, but Mayr completely ignores the question of how so many new structural types were produced rapidly in the Cambrian explosion. Once they formed, it may be hard to change them. But what process formed them originally? Mayr does not say.

Likewise, Mayr, in another place, writes: "The evidence indicates that the older a taxon is, the more difficult it is to escape from the straitjacket of a highly integrated (congealed) genotype. This is why not a single new morphological type (phylum) has originated since the Cambrian,

---

59. Gould, *Wonderful Life*, 24.
60. Mayr, *Toward a New Philosophy of Biology*, 434.

over 500 million years ago."[61] Mayr tacitly affirms that natural selection lacks the power to break the straitjacket imposed on organismal form by the cohesion (or congealing) of the genotype since it has been unable to create a new structural form over the last half billion years. So it should occur to him to ask whether natural selection can account for the original creation of those structural forms in the Cambrian explosion. But once again, Mayr simply ignores this most important of questions.

Mayr is not alone. The Darwinian apologist Jerry Coyne challenges his creationist foes with the words, "the fossil record gives no evidence for the creationist prediction that all species appear suddenly and then remain unchanged. Instead, forms of life appear in the record in evolutionary sequence, and then evolve and split."[62] This is technically true. The fossil record does not support strict creationist notions of all species popping into existence suddenly. But it also does not support Coyne's idea of life appearing in evolutionary sequence. The major structural forms of animal life do not appear in sequence but rather all together and rather suddenly in the Cambrian explosion, and then they remain virtually unchanged down to the present. Coyne's anticreationist polemic is a bit disingenuous. Later, Coyne provides a list of unanswered questions that evolutionary biologists continue to investigate. One of those questions, not surprisingly, is, "What caused the Cambrian 'explosion' of life, in which many new types of animals appeared within only a few million years?"[63] If Coyne is willing to acknowledge this as an unanswered question (despite his scare quotes around the word *explosion*), he cannot at the same time say "that the major tenets of Darwinism have been verified. Organisms evolved, they did so gradually, lineages split into different species from common ancestors, and natural selection is the major engine of adaptation."[64] Until the cause of the Cambrian explosion is known, the major tenets of Darwinism remain uniquely unverified! But such are the problems the Cambrian explosion causes those who are committed to the grand narrative of Darwinian triumph.

Even where biologists try to offer an alternative solution that "defuses" the explosion, they engage in a curious form of argument. In 1998, Lindell Bronham and his colleagues published a paper employing

---

61. Mayr, *One Long Argument*, 160.
62. Coyne, *Why Evolution Is True*, 32.
63. Ibid., 223.
64. Ibid.

molecular clock analysis to argue that a long period of gradual diversification of early animals must have preceded the Cambrian period, in what they called a "Precambrian phylogenetic fuse hypothesis." Since there was a long Precambrian fuse, there was no real explosion. But, curiously, in making this argument they write:

> Although we cannot provide precise estimates of the origin of metazoan [animal] phyla, we can use our results to confidently reject the Cambrian explosion hypothesis, *which rests on a literal interpretation of the fossil record* and assumes that special evolutionary phenomena, capable of producing profound differentiation in a short period, operated in the Cambrian but not before or since.[65]

Can one interpret the fossil record in any way other than literally? What would a metaphorical interpretation of fossils look like? Even more curiously, the authors continue: "By contrast, the Precambrian phylogenetic fuse hypothesis *assumes no more than we already know to be reasonable*: that lineages can diverge gradually over time and that the fossil record contains gaps that can greatly reduce the chances of finding fossils for certain periods or particular types of organisms."[66] As a subjective criterion, reasonableness is not a scientific category. What is reasonable to one person may be deemed entirely unreasonable to another. The authors do not appeal to "no more than we already know to be *true*," but to "no more than we already know to be *reasonable*." The authors may find a Darwinian explanation to be reasonable, but this explanation is not rendered true just by virtue of their subjective opinion that it is reasonable. Their ideological commitment to Darwinism is what forces them to reject the significance of the Cambrian explosion, not scientific evidence.

Most recently (in 2018), Daley and her colleagues try to downplay the non-Darwinian nature of their findings by calling the development of the euarthropoda a "gradual" event that played out over perhaps forty million years. But forty million years is less than 1 percent of the age of the earth, and given that no new animal phyla have evolved over the last 260 million years since the Permian extinction, the findings of Daley and her colleagues underscore the "sudden," not gradual, nature of the Cambrian explosion. Abundant evidence exists of suitable deposits for

---

65. Bronham et al., "Testing the Cambrian Explosion Hypothesis," 12388 (italics added).

66. Ibid. (Italics added).

fossil formation in the Precambrian period. Euarthropoda, therefore, must not have appeared until about 541 million years ago at the earliest. They conclude: "The fossil record of euarthropoda provides our most complete view of the origin and radiation of a major phylum during the Cambrian explosion. Rather than a sudden event, this diversification unfolded gradually over the ~40 million years of the lower to middle Cambrian, with no evidence of a deep Precambrian history."[67] Daley and her colleagues try to downplay the non-Darwinian nature of their findings by calling the development of the euarthropoda a "gradual" event that played out over perhaps forty million years. But forty million years is less than 1 percent of the age of the earth, and given that no new animal phyla have evolved over the last 260 million years since the Permian extinction, the findings of Daley and her colleagues underscore the "sudden," not gradual, nature of the Cambrian explosion.

One final entry on the topic of macroevolution deserves comment. In a 2016 book, *The Origin of Higher Taxa*, T. S. Kemp argues unequivocally against the idea that natural selection can account for large-scale evolutionary change.[68] Kemp observes that for all but a tiny handful of speciation events documented in the fossil record, no intermediate stages are known.[69] There is just little hard evidence that organisms diverged into forms constituting higher taxonomic categories through slow, gradual, microevolutionary change. The primary reason for this concerns a problem we have already met in earlier contexts: how can organisms evolve through the accumulation of incremental changes in just one or a few traits when such changes would likely denigrate the functionality of the organism? Kemp addresses this problem by postulating what he calls a "correlated progression model."

Kemp's model, in which all the characters of an organism are assumed to be functionally integrated, offers an alternative explanation to the Darwinian process that tends to treat characters as if they can vary in isolation. In the correlated progression model, "small changes occur in succession in many functionally correlated characters, which leads eventually to major evolutionary transitions while the integration is maintained throughout."[70] In some ways, Kemp's description of cor-

---

67. Daley et al., "Early Fossil Record of Euarthropoda."
68. Kemp, *Origin of Higher Taxa*, vii.
69. Ibid., 3.
70. Ibid., 36.

related progression merely states the obvious. We know that organisms must evolve in such a way that functionally correlated characters change together in harmony. But the question is, how does this occur when the variations on which natural selection acts are believed to occur randomly, not in a coordinated way? Before we take up this question, consider an example Kemp gives to demonstrate the importance of functional integrity in large-scale evolution.

Kemp considers the question of how ectotherms (cold-blooded organisms like snakes and lizards) evolved into endotherms (warm-blooded creatures like birds and mammals). What kinds of changes would be necessary to support an endothermic lifestyle? Kemp provides an answer in an extended paragraph that needs to be quoted in full:

> Endothermy is a far from simple system. At the cellular and molecular level, it requires a large increase in metabolic heat production, with a basal (resting) metabolic rate five to ten times that of a comparable sized ectotherm and a maximum aerobic rate increased by a similar ratio. This involved the evolution of increased numbers of larger mitochondria in the cells, and a shift in their biochemical activity to produce more heat directly, in place of ATP. But increased mitochondrial heat production alone is not enough. There must also be the means to balance heat production and heat loss, which requires as a minimum detection of body temperature, hormonal regulation of metabolic rate, variation in skin conductivity, finely controllable vascularization of blood flow to the body surface and oral cavity for evaporative cooling, and a whole variety of behavioral reactions, daily and seasonal, all precisely integrated by the interaction of the somatic and autonomic nervous systems. Beyond these, there are structures and mechanisms that evolved for enhanced ventilation rates, both resting and even more extremely during very high levels of aerobic muscular activity. To fuel the system, the rate of food assimilation needs to match the increase in metabolic rates, with implications for the sensory physiology of food detection, the mechanics of food collection and intake, the process of assimilation, and the need for large storage reserves in the body. All these activities and processes require an increased level of neuromuscular control.[71]

In short, the evolution of warm-bloodedness required an almost complete overhaul of all body systems. An organism receiving a variation

71. Ibid., 126.

producing more heat without at the same time receiving a variation allowing it to dissipate heat would quickly overheat and die. Given the complex systemic changes required for an endothermic lifestyle, it is hard to see how endothermy could evolve in a slow, stepwise fashion. For Kemp, the correlated progression model solves this problem. But does it?

Whereas Kemp argues for the obvious necessity of an organism maintaining its functional integrity during evolution, he admits that the developmental mechanisms that maintain this integrity are "quite poorly understood."[72] Further, he argues that "the genetic basis of correlated progression resides in the architecture of the genetic regulatory system, although little is yet known in detail about this."[73] Kemp's correlated progression model is much more a statement of the problem than a demonstrated solution. Organisms must be able to maintain functional integrity in order to evolve and diverge into higher taxonomic categories, but Kemp does not really know how this occurs. But, like Goldschmidt before him, he realizes it cannot occur by a strictly Darwinian process. After all, the fossil record does not support the Darwinian scenario. Kemp writes:

> The complexity of changes involved in the origin of a higher taxon, coupled with the relative dearth of actual evidence about its course, means that an account can always be forced into the straitjacket of microevolutionary processes. Suitable missing intermediate stages can be invoked and appropriate environmental circumstances imagined that complete the simple adaptive scenario.[74]

Here Kemp gets to the heart of the matter. Those who argue that a Darwinian microevolutionary process can account for the origin of higher taxa base their view on imagined scenarios, not documented evidence, such as when we are told that the bones of the reptilian jaw evolved into the bones of the mammalian inner ear. This is an example of the grand narrative of Darwinian triumph hard at work.

But try as they may, the supporters of the grand narrative cannot lay all evolutionary processes at the feet of Darwin. For there *is* a well-documented non-Darwinian method of "sudden" speciation at work in

---

72. Ibid., 58.
73. Ibid., 177.
74. Ibid., 169.

nature known as polyploidy. The existence of polyploidy is fully accepted by Darwinians, so how do they handle this inconvenient fact?

Polyploidy

In his *Genetics and the Origin of Species*, Theodosius Dobzhansky made the following comment at the beginning of a chapter on polyploidy:

> It has been pointed out in Chapter III that the sudden origin of a new species by gene mutation is an impossibility in practice. The argument employed to prove this thesis is simple enough. Races of a species, and to a still greater extent species of a genus, differ from each other in many genes, and usually also in the chromosome structure. A mutation that would catapult a new species into being must, therefore, involve simultaneous changes in many gene loci, and in addition some chromosomal reconstructions. With the known mutation rates the probability of such an event is negligible. The process of species formation is apparently a slow and gradual one, consuming time on at least a quasi-geological scale.
>
> It is highly remarkable, therefore, that alongside this slow method of species formation there should exist in nature a quite distinct mechanism causing rapid, sudden, cataclysmic emergence of new species.[75]

Highly remarkable indeed! After explaining how a Darwinian mechanism is unable to produce the sudden appearance of new species, Dobzhansky is forced to admit that there *is* a non-Darwinian process of speciation at work in the world of nature capable of producing the very thing he says a Darwinian process cannot—the "rapid, sudden, cataclysmic emergence of new species." In light of the reality of polyploidy, which George Ledyard Stebbins called "one of the most widespread and distinctive features of the higher plants,"[76] why is there so much dogmatic defense of natural selection as the primary (or even sole) mechanism of evolution?

Polyploidy refers to the well-documented phenomenon, best attested in plants, of a spontaneous doubling of the chromosome complement in the offspring of a mating. That is, instead of each parent sending half its chromosome complement to its offspring so that the offspring will have the same number of chromosomes as its parents, the offspring receives

75. Dobzhansky, *Genetics and the Origin of Species*, 192.
76. Stebbins, *Variation and Evolution in Plants*, 298.

the full complement of each parent's chromosomes, therefore ending up with twice the chromosome complement of its parents. It would be like a human baby being born with ninety-two chromosomes in each cell nucleus instead of the standard forty-six. Polyploidy comes in two varieties: *autopolyploidy* (doubling of the chromosomes in the offspring of parents of the same species) and *allopolyploidy* (doubling of the chromosomes in the offspring of a hybrid mating). Polyploid offspring are unable to effectively breed with representatives of the parent stock, meaning that polyploid individuals are considered unique, new species. Through the process of polyploidy, then, a new species can come into the world virtually overnight—in one generation. No gradual, Darwinian process is needed to produce these new species.

Just how prevalent is polyploidy? There is some uncertainty about this. In 1965, Ernst Mayr would echo Stebbins in saying, "Polyploidy is very widespread among plants and is one of the important mechanisms of speciation in the plant kingdom."[77] Further, he stated that "in the lumbricid earthworms . . . up to 70 percent of the species found in a given area may have arisen by polyploidy."[78] So according to Mayr, polyploidy was a major force in plant evolution and had been found in at least one kind of animal. But fifteen years later, he reversed himself: "In plants polyploidy can indeed lead to the instantaneous saltational origin of new species. However, this is presumably the minority process even in plants; in animals it is virtually unknown."[79] Why Mayr contradicts himself is unclear, but his 1965 assessment of the prevalence of polyploidy seems nearer to the truth. It is now generally accepted that polyploidy represents one of the principal evolutionary processes leading to speciation in plants, and it is also found in many forms of animals, possibly even in mammals.[80] Polyploidy is not a rare evolutionary process but a rather common one, a point of embarrassment for those who would hold to natural selection as the primary mechanism of evolution.

We can see this by once again looking at what Ernst Mayr has to say about polyploidy. Consider the following statements from various of his writings:

---

77. Mayr, *Animal Species and Evolution*, 440.
78. Ibid., 443.
79. Mayr, "Role of Systematics," 131.
80. Soltis, "What We Still Don't Know about Polyploidy."

- A single mutation does not make a new species except in the case of polyploidy. New species are due to the gradual accumulation and integration of small genetic differences.[81]
- With few exceptions (for example, polyploidy), every evolutionary phenomenon is simultaneously a genetic phenomenon and a populational one.[82]
- Evolution is a gradual process and, in general, so is the multiplication of species (except by polyploidy).[83]
- Very few isolating mechanisms are all-or-none devices. They are built up step by step (except in polyploidy).[84]
- Such a process (speciation through hybridization) has often been postulated, but has never been unequivocally established. The difficulty, except in the case of allopolyploidy, consists in keeping such a population of hybrids segregated until it has acquired reproductive isolation.[85]
- There is no mechanism (except instantaneous speciation through polyploidy) by which the cohesion might be loosened to the extent that it would permit the sympatric development of two independent gene complexes.[86]

Repeatedly, while trying to assure his readers that evolution works in a slow, gradual, Darwinian fashion, Mayr is forced to constantly refer to polyploidy as an exception. But as we know today, polyploidy is not a mere exception to the rule. In some plant groups and perhaps even in some animal groups, it is the rule. Natural selection is not the sole (and may not even be the primary) mechanism of evolution. Goldschmidt was not so far off base when he postulated chromosome repatterning as an important evolutionary mechanism.

Consider the words of ecologist Dolph Schluter:

> Speciation is one of the least understood major features of evolution. The main obstacle to progress is the variety of mechanisms

---

81. Mayr, *Systematics and the Origin of Species*, 225.
82. Mayr, *One Long Argument*, 138.
83. Mayr, *Animal Species and Evolution*, 24.
84. Ibid., 26.
85. Ibid., 130.
86. Ibid., 527.

that might lead to the evolution of reproductive isolation, any one of which can be difficult to rule out in a specific case. The upshot is that it is still difficult to point to even two species in nature and state with confidence the mechanism that produced them. The exceptions are speciation events resulting from polyploidy, because polyploidy leaves a clear genetic signature for a substantial period of time.[87]

Whereas the grand narrative of Darwinian triumph trumpets Darwinian natural selection as the primary mechanism of evolution, the reality is that non-Darwinian polyploidy is the only mechanism of evolution that has been well documented in nature; the rest is mostly speculation. Why have evolutionary biologists had such trouble embracing polyploidy? Why do they want to relegate it to the status of a mere exception to the Darwinian rule? Probably because polyploidy is somewhat mysterious. We really don't know how and why it occurs. What controls whether the hybrid offspring of two plants, for example, will have a normal set of chromosomes or a double set?

Polyploidy has been studied in recent years in a new species of polyploid plant known as *Tragopogon miscellus*, a yellow daisy-like flower that grows along roadsides in the Pacific Northwest. It is a hybrid formed by the mating of two different but closely related species that were transplanted from Europe. The two parent species hybridized in Europe, but the offspring were never fertile. But in America, a polyploidy hybrid was formed that was fertile and quickly spread. In a 2001 *ScienceDaily* article, Richard Buggs was quoted as saying, "What we found was a surprise. It's as if hybridization and chromosome doubling hit a reset button on gene expression, turning them all on—this could allow subsequent generations to experiment by switching off different genes."[88] In the same article, Pam Soltis says, "The expression of the hybrid plant's genes in all tissues at all times allowed natural selection to shape what would emerge generations later." But natural selection has no foresight, and therefore by definition cannot shape what will emerge in later generations. Both Buggs's and Soltis's comments raise the uncomfortable possibility (for Darwinians) that evolution through polyploidy may in fact have an ability to respond to environmental challenge in some intentional way.

At least that is what creationists would like to believe. In response to the *ScienceDaily* article about *Tragopogon miscellus*, the Institute for

87. Schluter, "Ecology and the Origin of Species," 378.
88. "Flowering Plant Study 'Catches Evolution in the Act,'"

Creation Research seized on this study, concluding that "according to these study authors, the real source of change in these daisies was not their environment, but their DNA. And their DNA contains coded instructions to rearrange itself in an organized way, so that future generations might be better equipped to survive in different environments. It is God's marvelous biological design—not evolution—that has been caught in the act."[89] Creationists see the doubling of the chromosomes as God's way to provide the resources whereby future generations of the polyploid plant will be able to rearrange their DNA in an organized and intentional way to respond to environmental challenge. Though we don't necessarily need to pose the action of God here, the teleological implications of Buggs's and Pam Soltis's comments cannot be so easily dismissed.

Christian Parisod and his collaborators have studied the role *transposable elements* play in polyploidy. Transposable elements (TEs) are genetic fragments that have the ability to move around from one place to another in the genome; they are sometimes referred to colloquially as "jumping genes." They were first discovered by the well-known corn geneticist Barbara McClintock in the 1950s, a discovery for which she won a Nobel Prize. We will consider McClintock's work in more detail in the next chapter. For now what is important is Parisod's argument that polyploidy seems to activate the movement of TEs, which allows for the reorganization of the genome.

> To summarize, the effect of allopolyploidy on TE genome fractions may be more complex than generally assumed. As TEs are abundant and disbursed throughout genomes, they predispose a young allopolyploid genome to rapid shuffling, participating in natural genetic engineering and producing abundant raw material for adaptive evolution at a crucial moment.[90]

Strikingly, both Parisod and the creationists invoke engineering here, though they obviously differ on the identity of the engineer. Either way, engineering is an intentional process. Polyploidy may challenge the sufficiency of Darwinian evolution not only by producing saltational speciation (perhaps Thomas Huxley was right when he chided Darwin for rejecting saltation so unreservedly!) but also by introducing the possibility of intentional, teleological evolution. As we will see in the next chapter, the evidence for "natural genetic engineering" as an evolutionary

---

89. Thomas, "Did Flower Study Catch Evolution in the Act?"
90. Parisod et al., "Impact of Transposable Elements," 44.

process is hard to ignore, much to the chagrin of the supporters of the grand narrative. If evolution is a teleological process, the whole religion–science relationship will need to be fundamentally rethought. At the very least, polyploidy vindicates the instincts of early Mendelians like Hugo de Vries and William Bateson that evolution may proceed in jumps, not always in a gradual, Darwinian way.

In the previous chapter and this one we have engaged several of the major works of the evolutionary synthesis, works that have become the heart of the modern evolutionary canon. From Dobzhansky to Simpson to Mayr to Rensch natural selection as the primary mechanism of evolution was assumed, and genetic, systematic, and paleontological data were interpreted through the lens of this assumption in order to show how natural selection *could* explain the data. But little evidence was presented requiring natural selection to be the only possible explanation, and the significance of contrary evidence was largely ignored. Goldschmidt refused to accept the growing Darwinian orthodoxy and was eventually marginalized and ridiculed. But the 1953 discovery of the structure of DNA by Francis Crick and James Watson, likely the greatest biological discovery of the twentieth century, opened up the potential to render a final verdict between Darwinian orthodoxy and Goldschmidt's heresy. The grand narrative would of course give the triumph to Darwin, but once again by ignoring the significance of crucial evidence and by dressing up ideology in the garb of science. The revolution in molecular biology, short of finally establishing Darwin's theory, would only serve to render evolution more mysterious than ever.

6

Reframing the DNA Revolution

NINETEEN FIFTY-THREE WAS A landmark year in the development of modern biology; American researcher James D. Watson and British researcher Francis Crick published an article in the journal *Nature* that would forever change the face of the biological sciences. Watson and Crick had discovered the structure of DNA, a molecule that many scientists at the time had already viewed as the secret to heredity. Watson and Crick theorized that DNA came in the form of a double helix: two strands of nucleotide bases wound around each other with the nucleotide bases on each strand bound to those on the other to hold the whole structure together. The nucleotide bases were four in number labeled with the letters A (adenine), C (cytosine), T (thymine), and G (guanine) such that genetic information could be understood as encoded in the precise sequence of these four bases. In one of the great understatements in the history of science, Watson and Crick wrote, "It has not escaped our notice that the specific pairing we have postulated immediately suggests a possible copying mechanism for the genetic material."[1] And this, of course, was the most important implication of this revolutionary discovery. Since the nucleotide bases making up one strand of the helix were precisely coordinated with those on the other strand, if the two strands could be pulled apart, each would become a template for the creation of additional strings. The structure of the DNA molecule allowed for replication of its

---

1. Watson and Crick, "Molecular Structure of Nucleic Acids," 737.

genetic content. The secret of heredity had essentially been solved, and the world of biology would never be the same.

Watson and Crick, of course, did not formulate this theory alone. While they would be awarded the Nobel Prize (along with Maurice Wilkins) for this discovery, many other scientists in the 1940s and early 1950s were working on the structure of DNA, and their work certainly informed Watson and Crick. Some of the more important contributors were Erwin Chargaff, Linus Pauling, and perhaps the most overlooked woman pioneer in this biological revolution, Rosalind Franklin, without whose x-ray diffraction photographs Watson and Crick almost certainly would not have succeeded.[2] Be that as it may, Watson and Crick have become the face of the molecular revolution that forms the basis of virtually the whole of the modern biological sciences.

Four years after announcing this momentous discovery, Francis Crick laid the foundation for the development of the field of molecular biology by pronouncing what he called the Central Dogma of that emerging field. Given that DNA appears to be a digital information storage system, Crick suggested that the genetic information stored in DNA could move only in one direction: from DNA out to RNA (the molecule that transcribes the information stored in DNA), and from RNA out to the synthesis of proteins, the building blocks of all biological structures. In no case, Crick said, can information flow backwards, from proteins and RNA back into DNA. In other words, DNA sequences constitute a static information storage system. DNA provides the instructions for building biological structures, but no process exists whereby biological structures outside the nucleus of the cell can reach back in and alter the DNA sequences of a cell in any intentional way. If this is true, Crick's Central Dogma of molecular biology essentially guarantees the truth of Darwinian evolution. Why?

Rival theories to Darwin's, including Lamarckism, vitalism, and Goldschmidt's "hopeful monsters," all seem to imply the ability of an organism, or an organism's cells, to somehow alter hereditary information to intentionally respond to environmental challenge. The Central Dogma prohibits this. No force outside the nucleus of a cell can alter the DNA of an organism. But DNA must change over time for evolution to occur, for if DNA does not change, then no new biological structures and forms could ever develop. The history of life, of course, documents that

2. For a concise discussion of Franklin's crucial role in the discovery of the structure of DNA, see des Jardins, *Madame Curie Complex*, 180–95.

such large-scale changes have occurred. So how does DNA change if no force can intentionally alter it? Changes must occur only as the result of random errors introduced during the process of DNA replication, or random mutations. The Central Dogma and Darwinian evolution fit together hand in glove, and the significance of this cannot be overstated, as recognized by the historian of the molecular revolution, Horace Freeland Judson:

> Crick's assertion, since amply borne out, was of fundamental significance: Lamarckism, or the hereditary passing on of characteristics that an organism has picked up or had changed by the action of the environment—whether thought of as the giraffe's neck, the Jew's intellectualism, the winter-hardiness of wheat treated by Stalin's geneticist Trofim Lysenko, or the esoteric collective unconscious postulated by Carl Jung—had been declared dead before, but here at last received its definitive postmortem and was buried, clearing the air.[3]

Noted molecular biologist Jacques Monod went even further:

> With that [the Central Dogma], and the understanding of the random physical basis of mutation that molecular biology has also provided, the mechanism of Darwinism is at last securely founded. And man has to understand that he is a mere accident. Not only is man not the center of creation; he is not even the heir to a sort of predetermined evolution that would have produced either man or something very like him in any case. It is not true that evolution is a law; it is just a phenomenon, which is quite different.[4]

Given the sweeping philosophical implications Monod draws here on the basis of Crick's Central Dogma, we might ask what evidence Crick had for this idea, and why he used the religiously loaded term *dogma* to describe it.

In the article in which Crick first pronounced his Central Dogma, he wrote:

> My own thinking (and that of my colleagues) is based on two general principles, which I shall call the Sequence Hypothesis and the Central Dogma. *The direct evidence for both of them is negligible*, but I have found them to be of great help in getting

---

3. Judson, *Eighth Day of Creation*, 192.
4. Cited in ibid., 192.

to grips with these very complex problems ... Their speculative nature is emphasized by their names.[5]

By his own admission, Crick did not base his Central Dogma on direct evidence, and so he was forced to admit that he was articulating a *dogmatic* position. About the Central Dogma he further admits, "This is by no means universally held—Sir Macfarlane Burnet, for example, does not subscribe to it—but many workers now think along these lines."[6] Sir Macfarlane Burnet was a Nobel Prize-winning physician, recipient of the Royal Medal and Copley Medal, recipient of an honorary doctorate from Cambridge University and the Order of Merit in addition to his knighthood. So it is clear that at least one highly decorated scientist did not find the Central Dogma convincing. Why then did Crick hold to it given his own admission that the direct evidence for it was negligible?

Crick, we must remember, was not trained as a biologist. He studied physics in college and was close to receiving his doctorate when World War II broke out and a German bomb destroyed his lab. He then worked for the British Admiralty during the war, but upon leaving the Admiralty in 1947 made a turn to biological questions. Why? According to Horace Judson, Crick told him it was because he was an atheist and wanted to "throw light into the remaining shadowy sanctuaries of vitalistic illusions."[7] Further, Crick reports:

> My own motives I never had any doubt about; I was very clear in my mind. Because when I decided to leave the Admiralty, when I was about thirty, then on the grounds that I knew so little anyway I might just as well go into anything I liked, I looked around for fields which would illuminate this particular point of view, against vitalism.[8]

James D. Watson reports about his colleague, "Neither was politics or religion of any concern. The latter was clearly an error of past generations, which Francis saw no reason to perpetuate."[9] Given Crick's clear atheistic motivations to pursue biology, his preference for the Central Dogma, despite the lack of evidence, is not hard to understand. Restricting information flow to only one direction (DNA to protein) undermines any attempt

---

5. Crick, "On Protein Synthesis," 152 (italics added).
6. Ibid., 153.
7. Judson, *Eighth Day of Creation*, 87.
8. Cited in ibid., 88.
9. Watson, *Double Helix*, 64.

to view evolution as having any kind of intentionality or direction, a view of evolution that would have been too close to religious sensibilities for Crick's comfort. The Central Dogma squares with a purely naturalistic, Darwinian understanding of evolution. Ironically, the religious term *dogma* in Central Dogma seems to have had an antireligious purpose for Crick. But it was still an appropriate term, for it seems the Central Dogma was indeed perpetuated on dogmatic rather than evidentiary grounds.

The Central Dogma, not surprisingly, has become a sacred article of faith in the Darwinian grand narrative. But is it true? Criticisms of this dogma were raised early on by Nobel Prize-winning geneticist Barbara McClintock and have not stopped up to the present time. In this chapter I will engage McClintock's often overlooked work, which will lead me to discuss a debate over directed mutation that remains unsettled even to this day. In light of evidence potentially undermining Crick's Central Dogma, it will then be interesting to critically engage the many anthropomorphic analogies that supporters of the grand narrative have developed over the years to explain how natural selection works. Despite their emphatic assertions to the contrary, Darwinians regularly employ the language of intentionality and agency in descriptions of how this supposedly nonteleological process operates. This phenomenon occurs far too often to be passed off as a mere Freudian slip. It is virtually impossible to talk about natural selection without invoking agency, a truth that might speak volumes about the true nature of the evolutionary process.

Barbara McClintock

Born in 1902, Barbara McClintock (d. 1992) was one of the most important scientists of the twentieth century. Among her many awards and accolades were her election in 1944 to the National Academy of Sciences and her 1983 Nobel Prize in Physiology or Medicine (the first woman to win this award unshared). After a brief teaching career at the University of Missouri in the 1930s, McClintock took up a full-time research position at the Cold Spring Harbor Laboratory on Long Island, a biological research station funded by the Carnegie Institution of Washington; this position she held for the balance of her career. What was so important about her research?

McClintock studied the genetics of maize (the common corn plant). She became the world's expert on corn genetics, her well-documented intellectual brilliance allowing her to see a significance in patterns in the coloring of corn kernals that others overlooked. As early as the 1940s, she began to challenge the dominant conceptual foundation of genetics by proposing that genes may not be fixed in place on chromosomes, but that they might be able to move around and relocate from one part of the genome to another. She called this idea transposition, and proposed that by moving around, certain genes might be responsible for controlling the actions of other genes. All cells have the same DNA in their nucleus, but cells nevertheless come to function in quite different ways. Some cells become skin cells, others become muscle cells, and still others bone cells, for example. While all cells have the same DNA, they must not all express the same genes at the same time. Genes must be turned off and on in various combinations in different cells to account for how cells with identical DNA come to function so distinctly. According to one of her biographers, "McClintock's own work was teaching her, over and over again, that the genetic apparatus is more labile and flexible than the central dogma allowed."[10] If gene expression must be regulated, there must be a regulatory process at work, and the Central Dogma really couldn't account for this.

It is fair to say that McClintock's ideas were deemed controversial by the biological establishment of her time. Her biographer Evelyn Fox Keller writes:

> The results McClintock reported in 1951 were totally at variance with the view of genetics that predominated. The biggest problem was, if genetic elements were subject to a system of regulation and control that involved their rearrangement, what meaning was then left to the notion of the gene as a fixed, unchanging unit of heredity? Central to neo-Darwinian theory was the premise that whatever genetic variation does occur is random, and McClintock reported genetic changes that were under the control of the organism. Such results just did not fit in the standard frame of analysis.[11]

Nathan Comfort concurs, calling the crux of McClintock's theory the idea that "transposition was nonrandom. The whole system was controlled by some as yet undetermined force that initiated coordinated mass-transpositions."[12] Comfort tells of a 1958 conference held at the

---

10. Keller, *Feeling for the Organism*, 173.
11. Ibid., 144.
12. Comfort, *Tangled Field*, 148.

University of Wisconsin–Madison attended by both McClintock and the population geneticist Sewall Wright. McClintock presented her controversial views, which included drawing a diagram on the blackboard. Wright then approached the board and promptly erased McClintock's drawing, replacing it with one of his own. "After two or three cycles of this, McClintock stormed out of the seminar room, vowing never to return to Madison while Wright was still alive."[13] McClintock was stepping on the toes of Darwinian orthodoxy and, like Goldschmidt, she felt the backlash.

Yet despite the threat McClintock's ideas posed to the Central Dogma and by extension to the whole Darwinian edifice of evolutionary theory, the biological establishment was eventually forced to accept the reality of what came to be called transposable elements, genetic fragments that can indeed move around in a genome. The evidence that transposable elements exist became overwhelming, and it was for this discovery that McClintock was eventually awarded the Nobel Prize. But while absorbing the reality of genetic transposition into standard genetic theory, the biological establishment simply ignored the more radical implications of McClintock's work. What mechanism controlled genetic regulation and directed the action of transposable elements? McClintock did not provide an answer to this most vexing question. That she did not propose a public answer to this question is why the community of biologists was able to simply ignore it.

Evidence suggests that McClintock had a strong mystical side to her personality, not shared by most of her scientific contemporaries. She showed interest in things like ESP, UFOs, Buddhist meditation, and the like. She may not have been beholden to a purely materialist view of reality, and so she might have been less troubled by the teleological implications of her work. Comfort writes, "Scientists often take plausibility as a quick and dirty test of a hypothesis. If they cannot think of a mechanism by which something might occur, they are unlikely to take it seriously." But he continues, "McClintock stopped believing that extrasensory perception could not exist simply because we knew of no mechanism for it, or that UFOs could not exist because we had never confirmed a sighting. Her corn showed her that nature was filled with new forces and unexplained behavior."[14] Science eschews mystery, but as we saw in the

13. Ibid., 171.
14. Ibid., 154.

previous chapter, evidence is mounting that the actions of transposable elements may play an important role in the phenomenon of polyploidy in a way we do not fully understand. McClintock's more radical ideas cannot be easily dismissed.

Barbara McClintock's Nobel Prize address was published in the journal *Science* under the title "The Significance of Responses of the Genome to Challenge." The teleological nature of this title is unmistakable, and the contents of the article do not disappoint. McClintock talks of genomes having the ability to reorganize themselves when faced with a challenge for which they are unprepared, and she suggests that rapid reorganizations of genomes (not unlike what had been proposed by Richard Goldschmidt) might be responsible for some species formations.[15] More specifically, cells, according to McClintock, have the ability to sense the presence of ruptured ends of chromosomes in their nuclei and to activate a mechanism designed to bring the ruptured ends together no matter how far apart they are. She comments: "The ability of a cell to sense these broken ends, to direct them toward each other, and to unite them so that the union of two DNA strands is correctly oriented, is a particularly revealing example of the sensitivity of cells to all that is going on within them."[16] McClintock admits that the sensing devices that initiate these processes of internal repair are "beyond our present ability to fathom." But a goal for future research would be to "determine the extent of knowledge the cell has of itself and how it utilizes this knowledge in a 'thoughtful' manner when challenged."[17] All this talk of cells and genomes possessing cognitive abilities will make Darwinians cringe, but undeterred McClintock finally concludes:

> In the future, attention undoubtedly will be centered on the genome, with greater appreciation of its significance as a highly sensitive organ of the cell that monitors genetic activities and corrects common errors, senses unusual and unexpected events, and responds to them, often by restructuring the genome. We know about the components of genomes that could be made available for such restructuring. We know nothing, however, about how the cell senses danger and instigates responses to it that often are truly remarkable.[18]

15. McClintock, "Significance of Responses," 793.
16. Ibid., 794.
17. Ibid., 798.
18. Ibid., 800.

McClintock ventures into territory far beyond the bounds of the Central Dogma and Darwinian orthodoxy. That we don't know exactly how these things happen allows Darwinians to ignore these possibilities. But this story does not end with Barbara McClintock. The mystery only deepened in 1988 when a paper appeared claiming to have demonstrated the reality of "directed mutation" in bacterial populations.

Directed Mutation

In 1943, Salvador Luria and Max Delbrück published a seminal paper seeming to close the door on any controversy over the random nature of the mutations on which natural selection acts. These two biologists grew bacterial colonies and plated them on an environment containing a selective agent (a virus) that would kill them. They then calculated how many bacteria on each plate had developed resistance to the virus. Luria and Delbrück theorized that if different plates contained about the same numbers of virus-resistant bacteria, then the mutation conferring resistance must have occurred after the bacteria were placed into the hostile environment and in response to the hostile environment. That is, the mutations must have been induced intentionally to adapt to the hostile environment. But if different plates contained differing amounts of virus-resistant bacteria, then the mutation must have occurred randomly before the bacteria were introduced to the hostile environment so that different plates received different numbers of mutant forms from the original colony. This became known as the fluctuation test. Luria and Delbrück found that the number of virus-resistant bacteria varied widely from one plate to another, seeming to confirm the idea that mutations are generated randomly.[19] This classic study became a bedrock of Darwinian evolution.

But things changed in 1988, when John Cairns, Julie Overbaugh, and Stephen Miller, researchers at the Harvard School of Public Health, published a paper purporting to demonstrate directed mutation. Cairns and his colleagues recognized the seminal nature of Luria and Delbrück's experiment but argued that the it had shown only that some mutations arise randomly, not necessarily that all do. By placing bacteria in an environment containing a deadly virus, only bacteria that had already received a random mutation conferring resistance would survive and

---

19. See Luria and Delbrück, "Mutations of Bacteria."

reproduce; others would be quickly killed by the virus before they could respond to it. But Cairns and his collaborators wondered what would happen if bacteria were introduced to a challenging but less harsh environment, giving the bacteria time to respond to the environmental challenge. Would they?

The Cairns team thus repeated Luria and Delbrück's fluctuation test, but instead of using a virus as the selective agent, they employed the sugar lactose. Bacteria metabolize sugars for energy, so the Cairns team grew a strain of bacteria lacking a gene for lactose metabolism. Once they were plated on a lactose environment, mutant forms that had already mutated in such a way as to now be able to metabolize lactose would begin growing right away. The lactose-resistant bacteria would not be able to grow, but they also would not die right away as was the case in Luria and Delbrück's virus environment. Given some time, would the lactose-resistant bacteria be able to generate a mutation allowing them to metabolize lactose and begin growing? Cairns, Overbaugh, and Miller report that they did, and saw this as unequivocal evidence for directed mutation:

> The main purpose of this paper is to show how insecure is our belief in the spontaneity (randomness) of most mutations. It seems to be a doctrine that has never been properly put to the test. We describe here a few experiments and some circumstantial evidence suggesting that bacteria can choose which mutations they should produce.[20]

We should not overlook the use of the words "belief" and "doctrine" here. The Cairns study, not surprisingly, touched off a major controversy in the biological literature full of such religious terms as "dogma," "orthodoxy," "heresy," and "blasphemy," strongly suggesting that this was not understood as merely a scientific controversy. But before considering the reaction in more detail, we must realize that Cairns and his colleagues were not the only ones making such startling claims for the reality of directed mutation.

At about the same time, Barry Hall published a pair of articles purporting to show what looked like anticipatory mutation. He placed *E. coli* bacteria into a challenging environment that would require the bacteria to receive two independent mutations in order to adapt to the more hostile situation. The first mutation in the sequence conferred no selective advantage on the bacteria. Only when the second mutation occurred

---

20. Cairns et al., "Origin of Mutants," 145.

was the challenge met. Hall's results showed that when exposed to the environmental challenge, the first (neutral) mutation occurred with frequencies orders of magnitude higher than would be expected if they were randomly generated, as if the bacteria were intentionally producing a large supply of these neutral mutations in anticipation of the production of the second mutation that would allow the bacteria to meet the environmental challenge. Hall wrote:

> If cells had a means of specifically increasing the rate of advantageous multiple mutations, they might be able to circumvent a barrier that would appear to be difficult when two independent random mutations are required to improve fitness and insuperable when more than two are required. Here I present evidence that *Escherichia coli* cells do, indeed, possess such a mechanism.[21]

This is a truly astonishing result. It implies the kind of cognitive abilities in cells that McClintock imagined. As Hall further says, "Because mutations are rare events, it is usually assumed that multiple mutations are the result of temporally separate, and therefore independent, events." But he believes this assumption needs to be critically reexamined.[22] What kind of response did Cairns and Hall elicit with their provocative work?

Neville Symonds, a British geneticist, was one of the first to respond. While he found Hall's work truly provocative, he argued for deferring acceptance of the conclusions since alternative explanations could be imagined: "As an alternative to Hall's interpretation, an explanation based on orthodoxy is that physiological conditions in old colonies induce a burst of random transposition activity in a fraction of the cells."[23] One might ask whether an "induced" activity can be considered random. But perhaps more important is the implication that Symond's alternative interpretation of Hall's results is driven by a desire to save "orthodoxy." Two years later, Symonds softened his tone somewhat, admitting that "biologists are slowly coming round to the idea that in certain circumstances bacteria may mutate in an apparently directed fashion."[24] More

---

21. Hall, "Adaptive Evolution That Requires Multiple Spontaneous Mutations: Mutations Involving Base Substitutions," 5883.

22. Hall, "Adaptive Evolution That Requires Multiple Spontaneous Mutations: Mutations Involving an Insertion Sequence," 887.

23. Symonds, "Anticipatory Mutagenesis?" 120.

24. Symonds, "Fitter Theory of Evolution?" 33.

importantly, religious terms continued to pile up in Symond's writing. Consider the following passages:

- At the heart of modern biology is a theory most of us take for granted: neo-Darwinism. Its central idea—that organisms evolve by a combination of random genetic change and natural selection—has become tantamount to *dogma*. So when respectable biologists turn up evidence which appears to run counter to the *creed* people naturally get excited. Splashy articles appear in newspapers. Commentators gather as though at the scene of a terrible accident, the running aground of the good ship scientific *orthodoxy*. Could He [Darwin] really have been wrong?[25]

- On the face of it, this [Cairns's work] was blatant *heresy*—how can an organism sense its needs in a particular environment and then mutate accordingly?[26]

- Once heralded a *blasphemy*, Lamarckian adaptations may eventually be seen as an inevitable consequence of molecular genetics.[27]

The evidence for directed mutation clearly threw biology into an ideological crisis. The evidence was too compelling to ignore. But the dogmatic nature of Darwinian evolution meant that as in the doctrinal disputes in the early church, orthodoxy had to be defended by branding opposing ideas as heretical or blasphemous rather than by simply showing them to be wrong. Symonds was certainly not alone in reflecting the ideological nature of the controversy over directed evolution.

In 1989, Richard Lenski, Montgomery Slatkin, and Francisco Ayala, took a much harsher tone toward Cairns and his colleagues. In their opinion, "if the hypothesis of directed mutation is, indeed, correct, it has onerous implications for bacterial genetics and, in particular, for the use of bacterial populations as model systems for the study of evolutionary processes."[28] But what would make Cairns's results onerous? If the directed mutation hypothesis is correct, would it not open up exciting new avenues of research into bacterial genetics and the study of evolution? Yes, but only if one is open to the possibility of a teleological view of evolution, something that will sound too religious to most biologists.

25. Ibid., 30 (italics added).
26. Ibid. (Italics added).
27. Ibid., 34 (italics added).
28. Lenski et al., "Mutation and Selection in Bacterial Populations," 2775.

The directed mutation hypothesis is onerous only to those with a vested interest in maintaining a strictly Darwinian view, a commitment that Lenski, Slatkin, and Ayala clearly share.

More interestingly they continue:

> Cairns et al. assert that directed mutation in bacteria (if it is demonstrated to exist) 'could, in effect, provide a mechanism for the inheritance of acquired characteristics.' We disagree with this claim; we also view it as potentially harmful in that it may seem to give credence to prescientific claims that have been thoroughly disproved.[29]

Lenski and his colleagues complain here about the "harmful" possibility that directed mutation could revive the ghost of what they label an empirically disproved, prescientific Lamarckism. But as we will see in next chapter, some modern biologists *are* reviving Lamarckian ideas in research on epigenetic inheritance; Lamarckian ideas are not outside the bounds of science, and there is nothing inherently harmful about them if one is not dogmatically committed to Darwinism. By labeling them "prescientific," Lenski, Slatkin, and Ayala seem to imply that evolutionary science somehow began with Darwin, an absurd idea. Lamarck was a recognized scientist even if one no longer accepts the truth of his work. Lenski, Slatkin, and Ayala could have simply charged Cairns et al. with giving credence to disproved ideas. But by labeling these ideas as "harmful" and "prescientific," they reveal the fear they harbor of a teleological view of evolution, which would cut the legs out from under a naturalistic Darwinism. This appears to be their real concern.

In another 1989 article, Lenski seeks alternative explanations for the results obtained both by Cairns and his colleagues and by Hall. He recognizes that "the mechanism proposed by Cairns et al., if correct, would radically alter our views of the molecular genetics of bacteria . . . Aren't there any simpler explanations?"[30] He then surveys a couple potential alternative explanations, explanations that "would probably not profoundly alter either molecular or population genetic orthodoxies," but he notes that none of these potential alternatives would explain Hall's results.[31] Lenski's overriding interest to maintain scientific orthodoxy is clear and is confirmed when after laying out the kinds of evidence that

---

29. Ibid., 2777.
30. Lenski, "Are Some Mutations Directed?" 149.
31. Ibid., 150.

would unequivocally prove directed mutation he concludes, "When such evidence has been gathered, we will know whether (and how greatly) we must alter current dogmas of molecular and population genetics."[32]

Four years later, Lenski and John Mittler emphasized that "the reality of directed mutation is disputed and its mechanistic basis, if real, is unknown."[33] But if such a mechanism were found, "it would provide an important exception to the neo-Darwinian tenet of random mutation, even if it operated only in certain micro-organisms."[34] Fortunately, in their view, no mechanism producing directed mutation has been shown to exist, and studies asserting that a directed mutation mechanism has been found have not excluded all possible alternative hypotheses based solely on random mutation.[35] But Lenski and Mittler are forced also to admit that directed mutation has received a great deal of attention because of its challenge to the core of Darwinian evolution, and because the evidence for it "has been published in major journals by several respected biologists and for many different experimental systems."[36] Lenski and Mittler, while wishing to explain the phenomenon away, recognize that they can't ignore the threat that directed mutation poses to the orthodox positions they hold so dear. The evidence for directed mutation is compelling.

In the same year the Lenski and Mittler article appeared, Donald MacPhee took on the issue of directed mutation in the pages of *American Scientist*. He begins: "We who teach biology tend to think that the basis of modern evolutionary theory is well established. As an article of *faith*, we accept the idea that organisms evolve by a combination of random genetic change and natural selection, the tenets laid out by Darwin."[37] Therefore: "Any notion that the right kinds of mutations arise in response to particular environmental stresses is antithetical to Darwinian *dogma* and is tantamount to arguing that the environment directs the evolution of new species."[38] The view put forth by Cairns, Hall, and others thus "contrasts sharply with the view of the scientific community in general;

---

32. Ibid.
33. Lenski and Mittler, "Directed Mutation Controversy," 188.
34. Ibid., 189.
35. Ibid., 193.
36. Ibid.
37. MacPhee, "Directed Evolution Reconsidered," 554 (italics added).
38. Ibid. (Italics added).

for some, it may even border on *heresy*."³⁹ So can Darwinian orthodoxy be saved? MacPhee thinks it can. His own research, he believes, provides an alternative explanation for what looks like directed mutation. When bacteria have glucose available as a food source, the genes for metabolizing other types of sugar are repressed in a process MacPhee calls catabolite repression. But when the bacteria are placed in an environment containing only lactose (as was the case in the Cairns group experiment), catabolite repression is lifted, and mutation rates increase dramatically, leading to the likelihood that a mutation will be randomly produced conferring on the bacteria the ability to metabolize lactose. He concludes, "In our experiments, catabolite repression, rather than 'directed evolution,' best explains the differential yields of lactose-positive and valine-resistant mutants."⁴⁰

Perhaps MacPhee has explained the evidence for "directed mutation" in a way as to save Darwinian orthodoxy, but before we declare the case closed, we should note that MacPhee begins the discussion of his alternative explanation with this passage:

> The model I prefer *assumes* that mutations arise continuously and spontaneously, in the best tradition of Darwin. The rates at which mutations appear are regulated but not directed in any way. Not only does this approach leave Darwinian tenets intact, but the genetic mechanisms for regulating mutation rates are also well within the framework of known metabolic activities for bacterial cells. In short, this model also spares the central dogma.⁴¹

He has started from the assumption of his preferred solution and interpreted the data through this lens. He even speaks of a "strong Darwinian bias in the scientific community,"⁴² a bias he clearly shares and which likely has influenced his interpretation of the results. We should also note that the phenomenon of catabolite repression is something MacPhee discovered while doing research that is quite different than the experiment of Cairns, Overbaugh, and Miller. So MacPhee has to write, "*Assuming* that such a system was operating during the Cairns experiment, it might be possible to reinterpret some of their conclusions."⁴³ All this talk of

39. Ibid. (Italics added).
40. Ibid., 561.
41. Ibid., 560 (italics added).
42. Ibid., 554.
43. Ibid., 560 (italics added).

Darwinian orthodoxy, bias, heresy, and the basing of scientific explanations on unsubstantiated assumptions certainly has the look of a field that is grasping at straws, trying to circle the wagons to save a sacred orthodoxy while at the same time being willing to sacrifice the pursuit of truth at the altar of this orthodoxy.

The controversy over directed mutation has not been resolved. Graham Bell (whose work on natural selection we analyzed in chapter 4) noted the controversy surrounding the work of the Cairns group and Hall, stating, "These experiments continue to be controversial, and a number of complicated but conventional processes that might explain them have been suggested."[44] Nothing more is said. But if one has to invent complicated processes to save conventional ideas, perhaps Occam's razor applies. Directed mutation *is* the simpler explanation, even if it fails to comport with orthodoxy. The situation had not gotten any clearer by 2014. In the *Princeton Guide to Evolution*, Charles Baer asked the question "How random is mutation?" He notes that evidence exists demonstrating how some bacteria do increase their mutation rates when under physiological stress. But, in his view, "it remains unclear whether the stress-induced increase in mutation represents an adaptation or is simply a feature of a sick organism functioning at a subpar level; evolutionary orthodoxy suggests the latter."[45] Perhaps. But orthodoxy is not necessarily truth.

Theodosius Dobzhansky, though writing well before the publication of experiments purporting to show directed mutation, certainly understood what was at stake: "But why, it may be asked, should useless mutations be produced at all? It would be vastly better if the organisms produced only useful mutations where and when needed. A little thought will show how naïve such a question really is."[46] Dobzhansky notes that normal colon bacteria produce large numbers of virus-resistant and antibiotic-resistant mutants, many of which are lost because they occur in a virus-free and antibiotic-free environment where they are not needed. But they are ready to flourish if a virus or antibiotic should appear in the environment. For bacteria to produce virus-resistant and antibiotic resistant mutants only in environments containing viruses or antibiotics,

---

44. Bell, *Selection*, 48.
45. Baer, "Mutation," 318.
46. Dobzhansky, *Evolution, Genetics, and Man*, 107.

"bacteria would have to possess a prescience of the future."[47] By the time a virus attacks, it is too late. So bacteria produce all kinds of mutants even if they are not immediately useful so that they may become useful if the environment changes. Dobzhansky concludes: "Herein lay the fundamental error of the belief in the inheritance of acquired traits; it assumed that the genes are made to mutate in a determined direction by every change in the development of the body."[48] In light of the work on directed mutation surveyed here, perhaps the error lay with Dobzhansky. He could not conceive of how directed mutation could be possible. How could bacteria have a prescience of the future? But Cairns, Hall, and others have provided evidence for this very possibility, a possibility that continues to be debated in the literature.

In a 2017 review article concerning the phenomenon of adaptive variation, Deborah Charlesworth, Nicholas H. Barton, and Brian Charlesworth conclude: "Natural selection can thus produce the appearance of directed mutagenesis. This model, while not fully confirmed experimentally, is consistent with all currently available data."[49] Perhaps the data of directed mutation *is* consistent with a Darwinian understanding, but it is also consistent with a teleological one. Consistency arguments do not constitute proof. That attempts to bring the data of directed mutation experiments into the Darwinian fold—attempts not yet receiving full experimental confirmation—continue thirty years after the original Cairns study is truly striking. The possibility of directed mutation cannot yet be disregarded. No one has done more to advance this idea and a whole new view of evolution based on it than the biologist whose characterization of evolution as a continuing mystery set the framework for this entire book—University of Chicago biologist James A. Shapiro.

Shapiro enjoyed a long working relationship with Barbara McClintock and came to accept some of her more radical proposals, which he laid out in detail in 2011 in *Evolution: A View from the 21st Century*. He asks, "How does novelty arise in evolution?" and declares, "Innovation, not selection, is the critical issue in evolutionary change."[50] He continues:

> In the context of earlier ideological debates about evolution, this insistence on randomness and accident is not surprising. It

47. Ibid.
48. Ibid.
49. Charlesworth et al., "Sources of Adaptive Variation," 8.
50. Shapiro, *Evolution*, 1.

springs from a determination in the 19th and 20th Centuries by biologists to reject the role of a supernatural agent in religious accounts of how diverse living organisms originated.[51]

The problem with maintaining this emphasis on randomness and accident, according to Shapiro, is that modern molecular biology shows that it simply isn't true. "The capacity of living organisms to alter their own heredity is undeniable. Our current ideas about evolution have to incorporate this basic fact of life."[52] Shapiro dubs this ability "natural genetic engineering" and marshals some impressive evidence to support it.

According to Shapiro, cells have an amazing ability to sense what is going on around them and react accordingly. They keep track of copying errors and repair them, adjust their metabolism to various nutrients, and interact appropriately with other cells. "Life requires cognition at all levels."[53] For example, *E. coli* reproduces its DNA with great precision (less than one mistake for every billion nucleotide bases), a fact that can only be accounted for if the cell can monitor the copying process and repair mistakes as they happen. Shapiro writes: "We can think of this two-level proofreading process as equivalent to a quality-control system in human manufacturing. Like human quality-control systems, it is based on surveillance and correction (cognitive processes) rather than mechanical precision."[54] And he is very explicit that he views this cognitive understanding of cells to be a direct challenge to Crick's Central Dogma.[55] Imbuing cells with the actions of mind (cognition) will not sit well with most Darwinians, but we should not forget that one of the principal figures of Darwinian evolution, Sewall Wright, held panpsychist views not unlike those held by Pierre Teilhard de Chardin. Shapiro is not the first biologist to view mind at work at lower levels of organization. But unlike Wright, he is open to the full radical significance of this.

Perhaps the easiest way to see cell cognition at work is in the immune system. When we are born, there is no telling what types of viruses and bacteria we will encounter during our lives. It is impossible to be born with the lymphocytes (immune system cells) necessary to counteract all possible microorganisms that we might encounter. Lymphocytes

---

51. Ibid.
52. Ibid., 2.
53. Ibid., 7.
54. Ibid., 14.
55. Ibid., 24.

therefore must be able to intentionally restructure themselves to counteract the specific micro-organisms we come in contact with. Shapiro warns: "We humans, for instance, could not survive if our lymphocytes (immune system cells) were incapable of restructuring certain regions of their genomes to generate the essential diversity of antibodies needed for adaptive immunity."[56] Traditional thinkers, according to Shapiro, argue that living cells cannot make adaptive use of their natural genetic engineering capabilities. They make this argument to protect their view of evolution as an undirected process based on random mutation. But this is a philosophical, not a scientific, position, and is utterly wrong. "The operation of a tightly regulated sequence of natural genetic engineering events in the adaptive immune system is probably the most elaborate example we have of purposeful genome manipulations."[57]

Shapiro continues to slay the sacred cows of Darwinism by viewing natural genetic engineering as a process that would allow for saltational evolution. Thomas Huxley chided Darwin for holding too closely to the doctrine of *natura non facit saltum*, and it appears he may have been right to do so. Shapiro states, "Contrary to the views of Linnaeus and Darwin, nature does indeed make leaps, and we now have molecular evidence of how some leaps occurred."[58] Protein evolution, according to Shapiro, does often proceed in large steps. And of course, he doesn't ignore the phenomenon of polyploidy, which we analyzed in the previous chapter: "Rapid plant evolution was "abominable" to Darwin because formation of interspecific hybrids and genome doubling are the kinds of sudden, genome-wide changes affecting multiple characters that he explicitly excluded from his gradualist, uniformitarian thinking."[59] People may have derided Goldschmidt and his hopeful monster theory, but he, along with the likes of de Vries, Bateson, and McClintock may have the last laugh.

Shapiro recognizes that the concept of natural genetic engineering may sound too much like the ID argument, one reason natural genetic engineering is resisted by many mainstream biologists. If there is engineering going on, there must be an engineer. Nevertheless, Shapiro thinks the evidence for natural genetic engineering is unequivocal. "Despite widespread philosophical prejudices, cells are now reasonably seen to operate

---

56. Ibid., 29.
57. Ibid., 56.
58. Ibid., 90.
59. Ibid., 122.

teleologically; their goals are survival, growth, and reproduction."[60] Shapiro recognizes just how controversial this idea will be:

> A shift from thinking about gradual selection of localized random changes to sudden genome restructuring by sensory network-influenced cell systems is a major conceptual change. It replaces the "invisible hands" of geological time and natural selection with cognitive networks and cellular functions for self-modification. The emphasis is systemic rather than atomistic and information-based rather than stochastic.[61]

How was Shapiro's book received? It is hard to say; it does not seem to have been widely reviewed. Perhaps most mainstream Darwinians just did not know what to say and instead ignored it, hoping it would garner little attention. But a few reviews did appear.

David Penny, while appreciating the wealth of molecular biological information in Shapiro's book, believes his thesis ultimately fails because it provides a non-Darwinian process for evolution ("'natural genetic engineering' is a catch cry throughout the book").[62] Though he does concede that this failure is a noble failure: "Noble, because it has so much information from molecular biology that is relevant to evolution. A failure because it does not specify any new grand mechanisms for evolution that could be tested."[63] As we have seen, this same criticism can be leveled at natural selection. Experimentally documenting the action of natural selection in the wild, and especially showing that natural selection accounts for large-scale evolutionary change, is still very much a hope, not a reality.

The lack of a clear mechanism for natural genetic engineering bothered another reviewer who is even more direct about the problem. Adam Wilkins wants to know the identity of the engineer in the phrase "natural genetic engineering." Can a cell be its own engineer?[64] By ignoring the role of natural selection in evolution, Wilkins finds it hard to imagine what force could be responsible for long-term evolutionary trends documented in the paleontological record, unless of course, we posit "supernatural or mystical forces such as the long popular but ultimately discredited force

---

60. Ibid., 137.
61. Ibid., 145.
62. Penny, Review of *Evolution*, 709.
63. Ibid.
64. Wilkins, Review of *Evolution* 424.

of 'orthogenesis.'"[65] Even if natural genetic engineering occurs at some level, Wilkins argues that these cellular processes themselves would have to have come into existence through natural selection, because "the only alternative for the origination of these capabilities, if one discards natural selection as the generative agent, is some supernatural force, a position I am certain is not being advocated here."[66] We see here the reappearance of the argument by default encountered in the work of population geneticist Ronald Fisher. Natural selection is the only *scientific* explanation for evolution. If it is wrong, a religious view is the only alternative, so natural selection wins by default. But this is not a serious scientific argument.

Not all reviews were negative. Bryn Bridges begins her review with, "I have been waiting more than thirty years for this book."[67] She applauds the book for challenging (if not completely undermining) the Central Dogma, and for making teleological thinking respectable. "Cognitive, goal-oriented processes do not have to be relegated to the realms of unscientific fancy and religion."[68] Shapiro's book comes with her highest recommendation. Wim Gaastra, a Dutch veterinarian, agrees, arguing that this book

> will increase your admiration for the ingenuity of the "genome" and its many possibilities for natural genetic engineering in a normal cell cycle. Your view on evolution as a process of random accidentally occurring mutations of which only a few survive, will have changed to evolution as an active process in which the genome uses all its available tools in acquiring new useful genetic information from outside its own cell.[69]

Evidence for natural genetic engineering continues to accumulate, even in the work of biologists who try to ignore its radical implications. French researchers have recently considered the impact of transposable elements on the evolution of the genome, citing earlier work that demonstrates how a specific transposable element (called the *Accord* LTR insertion) occurring in the common fruit fly (*Drosophila*) seems to confer protection against insecticides. They conclude: "Interestingly, these laboratory investigations are linked to natural population studies where

65. Ibid., 425.
66. Ibid., 426.
67. Bridges, Review of *Evolution*, 99.
68. Ibid.
69. Gaastra, Review of *Evolution*, 225.

fly populations collected around the world displayed a straight correlation between resistance and *Accord* LTR insertion."[70] If flies from around the world repeatedly undergo the same adaptive genome rearrangement when exposed to insecticides, this genome rearrangement cannot be occurring randomly but reflects the ability of the fly genome to respond directly and appropriately to environmental challenge. Commenting on this same phenomenon, Henry Chung and his colleagues had earlier called it "remarkable,"[71] and highlighted "the potential importance of [transposable elements] in rapidly providing the suitable molecular changes upon which selection can act."[72] Any talk of "suitable" molecular changes, of course, undermines any notion of randomness. Chung and his colleagues are right to comment on the "remarkable" nature of this phenomenon, even if they don't fully open themselves to its truly radical implication of cellular cognition.

The concept of directed mutation, like so many other areas of evolutionary theory, clearly remains a site of contestation. It has not been proved unequivocally to exist, but it has not been disproven either. But clearly, those who would like to disprove it are motivated by an ideological commitment to Darwinism; their criticisms must, therefore, be read in this light. The biological establishment will always resist explanations that seem to introduce teleology into the evolutionary process, an ironic circumstance given that members of this establishment have been introducing teleological language into evolutionary theory virtually since its inception. The scientific literature is shot through with anthropomorphic analogies and teleological language, a phenomenon that has the effect of deconstructing the supposedly nonteleological theories that are being described in such teleological terms. This should not surprise us. According to a quote widely attributed to J. B. S. Haldane: "Teleology is like a mistress to a biologist: he cannot live without her but he's unwilling to be seen with her in public."[73] Let's consider a few examples of "teleological Darwinism" in action.

---

70. Chénais et al., "Impact of Transposable Elements," 12.

71. Chung et al., "*Cis*-regulatory Elements," 1075.

72. Ibid., 1076.

73. The attribution of this quote to Haldane is uncertain. Some ascribe it originally to a nineteenth-century German physiologist E. von Bruecke. I take the quote from Deacon, *Incomplete Nature*, 107, where it is ascribed to Haldane without any sourcing to Haldane's writings.

Teleological Darwinism

Perhaps Darwinian evolution would be more convincing if its advocates did not continually employ teleological language even while explicitly rejecting teleological ideas. We find this happening when Darwinian proponents, casting about for analogies to describe the action of natural selection, employ comparisons that are entirely anthropomorphic and teleological. We find it also in the very language used to describe natural selection as well as the actions of cellular and molecular systems. Examples of this phenomena abound, and only a few can be considered in detail here.

Let's first consider analogies used to describe the action of natural selection. Darwin himself was the first to employ such an analogy when he compared natural selection to artificial selection, or what he called variation under domestication. Darwin was impressed by the significant variations human breeders could produce in everything from dogs to cattle to pigeons to plants in a relatively short period of time, and he theorized that given the much longer spans of evolutionary time, natural selection could surely produce even bigger variations. But Darwin seems to have ignored the major difference between artificial and natural selection: artificial selection is teleological. He seems to implicitly recognize this but ignores the larger significance:

> We cannot suppose that all the breeds were suddenly produced as perfect and as useful as we now see them, indeed, in many cases, we know that this has not been their history. The key is man's power of accumulative selection: nature gives successive variations; man adds them up in certain directions useful to him.[74]

Natural selection, of course, does not add up variations in a predetermined direction to produce a predetermined result. If accumulative selection is necessary to produce the vast variety of dog breeds, for example, perhaps Darwin might have considered whether accumulative selection would also be necessary to produce the far greater variations documented by the history of life on Earth.

Ernst Mayr makes this same error when commenting on the rejection of natural selection by the turn-of-the-century Mendelians:

---

74. Darwin, *On the Origin of Species*, 29.

> The various attempts to refute the power of selection were truly a comedy of errors. Bateson tried to do so (on his Russian trips) by establishing a correlation between phenotype and environment . . . None of those who denied the efficacy of selection so insistently bothered to visit the animal breeders, who at this very same period made constant advances in the productivity of their flocks and herds, owing to *deliberate selection*.[75]

Deliberate selection is teleological, natural selection is not. The error seems to lie with Mayr, not with Bateson and the other Mendelians. Artificial selection simply makes a poor analogy for natural selection, as the codeveloper of this theory, Alfred Russel Wallace, recognized: "No inferences as to variations in a state of nature can be deduced from the observation of those occurring among domestic animals . . . Domestic animals are abnormal, irregular, artificial; they are subject to variations which never occur and never can occur in a state of nature."[76]

Artificial selection aside, many other Darwinians have offered their own analogies to demonstrate the action of natural selection, with equally ironic results. Consider August Weismann:

> Those who agree with me in rejecting the *Lamarckian principle* will regard selection as the only *guiding* factor in evolution, which creates what is new out of the transmissible variations, by ordering and arranging these, selecting them in relation to their number and size, as the architect does his building-stones so that a particular style must result.[77]

Natural selection acting like an architect arranging building-stones to produce a particular result?! Weismann's characterization of the work of the architect is right on point. Architects do work from a preconceived plan to produce a particular style of building. If Weismann is serious in comparing natural selection to the work of an architect, then natural selection would be a highly teleological process. As an ardent Darwinian, Weismann certainly did not mean to indicate this. Nevertheless, it seems he could not resist smuggling teleology in the side door via an anthropomorphic analogy. George Ledyard Stebbins makes a similar move:

> But just as a sculptor creates a statue by removing chips from an amorphous block of marble, so natural selection creates new

---

75. Mayr, *Evolution and the Diversity of Life*, 349 (italics added).
76. Wallace, "On the Tendency of Varieties," 61.
77. Weismann, "Selection Theory," 65 (italics original).

systems of adaptation to the environment by eliminating all but the favorable gene combinations out of the enormous diversity of random variants which could otherwise exist.[78]

The sculptor, like the architect of course, has an end goal in mind. Natural selection does not.

Perhaps Ernst Mayr can come up with a better analogy:

> Why are modern motor cars so strikingly better than those of 75 years ago? Because all manufacturers constantly experimented with various innovations, while competition through customer demands led to enormous selection pressure. Neither in the automobile industry nor in the world of life do we find any finalistic forces at work, nor any mechanistic determinism.[79]

This is a curious comparison. Finalistic forces and mechanistic determinism are exactly what we find in the automotive industry. Automotive engineers do not randomly produce thousands of variable prototypes and submit them to the market to see which one will sell. They make intentional coordinated alterations to a single prototype and then enlist the marketing department to manipulate consumers into believing that the latest model is superior. Automobiles have improved over time precisely because of the highly teleological work of automotive engineers, who have a preconceived plan of the kind of car they want to produce and they think will meet customer demand. Consumers may at times provide a negative selective force, causing a particular car model to go extinct (remember the Edsel!), but overall, automotive engineering makes a poor analogy for natural selection, a point recognized by J. C. Greene: "Mayr analogized improvements in organisms to improvements in the motor car, projecting technological language onto the evolutionary process and thereby introducing a cryptic teleology into the description of evolution."[80] Whether natural selection is compared to a dog breeder, an architect, or an automotive engineer, this attempt to liken natural selection to a kind of familiar human work always ends with a teleological connotation. Are there any analogies that avoid this?

Francisco Ayala is one Darwinian who has embraced a limited teleology in evolutionary theory. He states:

---

78. Stebbins, *Variation and Evolution in Plants*, 104.
79. Mayr, *Toward a New Philosophy of Biology*, 259.
80. Greene, "History of Ideas Revisited," 222.

> Natural selection can be said to be a teleological process in two ways. Firstly, natural selection is a mechanistic end-directed process which results in increased reproductive efficiency. Reproductive fitness can, then, be said to be the end result or goal of natural selection. Secondly, natural selection is teleological in the sense that it produces and maintains end-directed organs and processes, when the function or end-state served by the organ or process contributes to the reproductive fitness of the organism.[81]

In a limited sense, natural selection can be understood as goal-directed if the goal is to produce greater reproductive fitness in an organism, since increased reproductive activity is what leads to the preservation of certain adaptive traits. But Ayala is very clear that "the process of natural selection is not at all teleological in a different sense. Natural selection does not tend in any way towards the production of specific kinds of organisms or towards organisms having certain specific properties."[82] But Ayala's distinction between these two senses of the word *teleology* seems undercut by the analogy he then employs to demonstrate the distinction:

> The creative process of natural selection must not be understood in the sense of the "absolute" creation that traditional Christian theology predicates of the Divine act by which the universe was brought into being *ex nihilo*. Natural selection may be compared rather to a painter which creates a picture by mixing and distributing pigments in various ways over the canvas. The canvas and the pigments are not created by the artist but the painting is.[83]

Of course an artist does not begin to paint without a preconceived idea of what kind of image she will paint. Brushstrokes are not applied randomly to many canvases with only one of the canvases being "selected" for further brushstrokes. Even in the case of abstract art, the artist knows in advance that she will paint an abstract piece rather than a nature scene, portrait, or still life. If natural selection acts like an artist painting a canvas, then natural selection would be teleological in the third sense discussed by Ayala (functioning to produce a specific result), despite his protestations to the contrary.

Perhaps the most famous attempt to find an analogy from human experience for natural selection came from the French molecular biologist

---

81. Ayala, "Teleological Explanations in Evolutionary Biology," 10.
82. Ibid.
83. Ibid., 5.

François Jacob. In his widely cited article "Evolution and Tinkering," Jacob likened natural selection to the work of a tinkerer bumbling about in his garage trying to fashion some kind of contraption from the materials he finds lying around. Natural selection does not work like an engineer, we are told, because an engineer works from a preconceived plan and has access to materials designed specifically to bring the plan to fruition.[84] Instead, natural selection works like a tinkerer who "does not know exactly what he is going to produce but uses whatever he finds around him whether it be pieces of string, fragments of wood, or old cardboards; in short it works like a tinkerer who uses everything he has at his disposal to produce some kind of workable object."[85] But workable for what? Even the tinkerer has an idea in mind of what kind of end product he is trying to fashion through his tinkering activity. When my daughter was in ninth grade, she was tasked by her science teacher to create a small vehicle that could be powered by air escaping a balloon. She made her vehicle out of scraps of chicken wire, milk straws, and old CDs for the wheels. This was tinkering at its best. Materials designed for quite other purposes were repurposed by my daughter, but always in service to the particular end goal of producing a working vehicle. Tinkering is a teleological process and therefore would seem to make a poor analogy for natural selection.

Jacob seems to recognize this. For before introducing the analogy of the tinkerer, he does admit that "natural selection has no analogy with any aspect of human behavior."[86] But even this admission does not prevent him from developing the tinkering analogy in an even more problematic way. For example, Jacob observes the following difference between tinkering and engineering: "When different engineers tackle the same problem, they are likely to end up with very nearly the same solution: all cars look alike, as do all cameras and all fountain pens. In contrast, different tinkerers interested in the same problem will reach different solutions, depending on the opportunities available to each of them."[87] This observation is further evidence for Jacob that natural selection works more like a tinkerer than an engineer since natural selection has produced an amazing variety of organisms. But Jacob seems to ignore here the well-documented phenomenon of convergent evolution that we

---

84. Jacob, "Evolution and Tinkering," 1163.
85. Ibid.
86. Ibid.
87. Jacob, *Possible and the Actual*, 35.

discussed in chapter 2. Like engineers tackling the same problem, evolution, when faced with similar environmental circumstances, *does* seem to converge on a limited number of similar solutions: so natural selection behaves more like an engineer than a tinkerer. Jacob's argument seems to anticipate a teleological view of evolution, despite his protestations to the contrary.

It is important in this context to recall the comment by Ernst Mayr in response to the phenomenon of convergent evolution. First, we should point out that Mayr explicitly agreed with Jacob on his tinkerer analogy: "As Jacob has said so rightly: 'Natural selection does not work like an engineer. It works like a tinkerer.'"[88] But when commenting on the phenomenon of convergent evolution, Mayr surprisingly wrote, "Convergence illustrates beautifully how selection is able to make use of the intrinsic variability of organisms to *engineer* adapted types for almost any kind of environmental niche."[89] Even Mayr seemed to implicitly recognize how the phenomenon of evolutionary convergence undermines Jacob's tinkerer analogy. If it is engineers, not tinkerers, who produce similar outcomes, then the engineer does seem to be a better analogy for the actions of natural selection, as James Shapiro argues in his own response to Jacob: "That is why the term *engineering* seems to be more appropriate for the built-in processes of self-modification that have operated over the course of evolution."[90]

One additional analogical example deserves notice. In chapter 4, we observed how Graham Bell employed the analogy of a (teleological) word game to demonstrate the action of natural selection. Others have followed with similar kinds of examples with similar problematic results. For example, David Barash compares natural selection to a process whereby we attempt to create the phrase "to be or not to be" via random substitutions of the twenty-six letters of the English alphabet. The odds that the first letter chosen will be a *t* is one in twenty-seven (including a space along with the twenty-six possible letters). The chance of then getting an *o* in the second slot would be one in twenty-seven times one in twenty-seven, or one in 729. The chance then of getting all eighteen characters right would be one in twenty-seven multiplied by itself eighteen times, an inconceivably small probability. But, Barash argues, what if

---

88. Mayr, *Toward a New Philosophy of Biology*, 153.
89. Mayr, *What Evolution Is*, 223 (italics added).
90. Shapiro, *Evolution*, 132.

instead of tossing out every meaningless letter combination and starting from scratch each time, we evaluate the "fitness" of a particular letter combination against the target phrase?

In one test, Barash says he began with "fuwl sazgh ekm fje." After several runs it had become "tubl hot nnoq ioby" and by run 22 it was "tub ep ok not ts e." By run 29, he had produced "to be ok not to bo," something looking quite similar to "to be or not to be." Barash concludes: "Starting with gobbledygook, and using only random variation and selective retention, something new had been created, something so nonrandom, in fact, that it is perhaps the most famous phrase in the English language!"[91] But of course, Barash's analogy fails. Natural selection does not evaluate the fitness of a variation against the template of a final target. It evaluates fitness against and only against current environmental realities. In a better analogy, we would require each phrase produced by random substitution of letters to be "fit" at each stage—meaning that only letter combinations that make sense in English would be preserved. Gobbledygook would be selected out. Given that the vast majority of random letter combinations would produce gobbledygook, the possibility of getting to "to be or not to be" via a series of steps of only meaningful phrases would be extremely remote. To make the analogy work, Barash has to smuggle in a teleological rule, but in doing so he completely undermines the whole purpose of the analogy—to demonstrate that a nonteleological process can produce something nonrandom!

Perhaps Jacob was right, and natural selection holds nothing in common with human behavior. But this would make natural selection a completely heterogeneous process, one capable of producing design in nature without any forethought or preplanning. This is exactly what Francisco Ayala argues when he characterizes Darwin's greatest discovery as the idea that one can get design without having to posit a designer.[92] But in our human experience, anything that is designed has a designer standing behind it. So can a nonteleological process really produce design? Darwinians are convinced that it can. Yet they are not happy to simply say so. They feel compelled to find a way to compare natural selection with some type of human activity, but they do so at the cost of introducing teleology into what they claim to be a nonteleological process. Why they so often do this one can only guess, but it is a phenomenon that

91. Barash, *Natural Selections*, 24.
92. Ayala, "Darwin's Greatest Discovery."

does not stop with the attempt to find an analogy for natural selection. Darwinian proponents speak about natural selection in teleologically loaded, anthropomorphic language all the time. Space will permit only a few brief examples.

One trope occurring regularly throughout the biological literature emphasizes the opportunism of natural selection. According to Ernst Mayr, "Life-forms have an astonishing capacity to vary, to respond to natural selection, and to take advantage of ecological opportunities."[93] When discussing eye evolution, Salvini-Plawen and Mayr state, "Natural selection, as always, is opportunistic and takes hold of whatever variable raw material it finds to guide it on to greater efficiency."[94] When it comes to coevolutionary relationships, Mayr tells us that "prey may develop better escape mechanisms that force the predator to improve its attack capacity," or that "flowers evolve in such a way as to make pollination more successful."[95] Examples like this of imputing agency to the evolutionary process could be multiplied many times over. Natural selection, of course, is not an active process but simply a term used to describe the results of differential reproduction in a genetically variant population of organisms. It therefore cannot "respond to," "take hold of," or "guide" anything. Nor can it "force" a predator to do anything or "cause" a flower to evolve in any specific way. The ease with which evolutionary biologists anthropomorphize the evolutionary process is truly striking.

George Gaylord Simpson recognized the problem and warned that *opportunism* was a dangerous word to be used in discussions of evolution because it carries a connotation of conscious action. According to Simpson:

> When a word such as *opportunism* is used, the reader should not read into it any personal meaning or anthropomorphic implication. No conscious seizing of opportunities is here meant, nor even an unconscious sensing of an outcome. The word is only a convenient label for these tendencies in evolution: that what can happen usually does happen; changes occur as they may and not as would be hypothetically best; and the course of evolution follows opportunity rather than plan.[96]

---

93. Mayr, *What Evolution Is*, 156.

94. Salvini-Plawen and Mayr, "On the Evolution of Photoreceptors and Eyes," 252.

95. Mayr, *What Evolution Is*, 210.

96. Simpson, *Meaning of Evolution*, 160.

But not everything that can happen does happen. Convergent evolution suggests that evolution is constrained, and that only a limited number of possible solutions are repeatedly realized. Only certain opportunities are ever "seized" upon by natural selection. Why constantly use the word *opportunism* if it is necessary then to drain it of its teleological connotations? It would be simpler to just not use the word at all. But to talk about natural selection in its true nonteleological manner simply fails to do justice to the amazing things the evolutionary process has been able to accomplish. Darwinians are addicted to teleological language for good reason, even if they refuse to acknowledge its larger implications.

One final example of the tendency toward teleological language is truly startling, especially in light of James Shapiro's work. One of the fundamental questions that evolutionary theory must answer is how the digital information code stored in DNA evolved. We now know that DNA is made up of triplets of nucleotide bases, with each combination of three nucleotide bases coding for a specific amino acid. For example, the DNA sequence CTC codes for the synthesis of the amino acid leucine, whereas the combination GTT codes for valine. Each three-nucleotide series is called a *codon*, and the different codon combinations are mapped onto the twenty different amino acids (several codons code for the same amino acid), which then get strung together to make protein molecules. But how did the codon/amino acid map develop in the first place? How did CTC come to mean leucine rather than valine?

In 1965, C. R. Woese proposed a theory to explain the development of the genetic code. He argued that early on in the history of life, cells would have functioned without a clear codon/amino acid map, and that these primitive cells would have been liable to a large number of translation errors as the DNA code was translated into amino acid sequences. Woese believes the genetic code developed as a result of primitive cells trying to improve their translation efficiency. The first thing to observe about Woese's argument is its highly speculative nature:

> However, what we shall <u>assume</u> about a very primitive translation system is that the mechanism was such that one could consider particular amino acids to be "assigned" to particular codons—it seems likely that *ambiguous* codon assignments <u>would have been</u> common, however. We further <u>assume</u> the mechanism was such that errors in translation were *extreme*— to such an extent that the probability of translating any given gene (or mRNA) entirely correctly was essentially zero. Finally, it is <u>assumed</u> that the pattern of errors characterizing the early

translation process was qualitatively like the above-described one, known to exist today.[97]

Woese has to begin by making several unsubstantiated assumptions about the workings of primitive cells. But even more striking is the language he then uses to describe how these primitive cells addressed the problem of poor translation efficiency:

> The primitive cell was faced with the seeming paradox that in order to develop a more accurate translation apparatus, it had first to translate more accurately. The way out of this paradox, of course, is that although unable *really* to reduce the translation error rate, the primitive cell can do something tantamount to this by <u>adjusting</u> the codon catalogue so that the *effect* of translation errors is lessened.[98]

Woese imputes agency to the primitive cell by imagining it to have the ability to recognize a paradox and respond to it by *adjusting* its codon catalogue. And the teleological language does not end there:

> Therefore, if the primitive cell started with a completely random, highly ambiguous set of codon assignments, the first step in improving the translation efficiency would be the gravitation of the codon assignments for the "functional" groups of amino acids toward the *least* error-prone codons . . . Once translation was improved to some extent by this <u>maneuver</u>, it might eventually become feasible for the cell <u>to make finer distinctions</u> within amino acids groups, <u>dividing</u> them into subgroups and so on, until finally, individual amino acids <u>would be recognized</u> as such.[99]

Woese imagines a primitive cell to have powers of discrimination and purposeful manipulation, the very powers James Shapiro actually does impute to cells, but an idea anathema to Darwinian orthodoxy. Ernst Mayr flatly stated, "It is illegitimate to describe evolutionary processes or trends as goal-directed (teleological)."[100] Despite all this teleological language, Woese seems not to want to take it literally. He concludes that his theory shows how primitive cells, "from the 'necessity' of minimiz-

---

97. Woese, "On the Evolution of the Genetic Code," 1548 (italics original, underlining added).

98. Ibid., 1549 (italics original, underlining added).

99. Ibid., 1550 (underlining added).

100. Mayr, *Evolution and the Diversity of Life*, 403.

ing the effects of translation errors, can evolve the highly ordered codon catalogue we observe today."[101] But Woese has not provided anything close to a Darwinian explanation of how the genetic code originally formed. Perhaps this is because it is impossible to imagine how such a code could form without the action of a cell imbued with some sort of primitive, but nevertheless, meaningful cognitive ability. Darwinians, of course, will balk at such a heretical suggestion, but their own language betrays them at every turn.

By laying bare the physical basis of heredity and genetic variation, the molecular revolution of the second half of the twentieth century should have firmly established natural selection as the primary driver of the evolutionary process. But instead it only deepened the mystery. With the discovery that DNA acts as a digital information storage system that maps onto amino acids in the form of a precise genetic code, new, still unanswered, questions were raised about how a process of random mutation and natural selection could possibly produce such a complex information system, and those who held to such an explanation were clearly motivated by strong ideological commitments. Further, evidence accumulated that DNA may not be a static storage system, but that the genome might actually be a dynamic entity capable of cognitive functions normally associated with mind and intelligence. From the 1960s on, numerous biologists have questioned the sufficiency of the Darwinian framework of evolutionary theory and proposed either new or extended theories to make sense of the newer data. Darwinism to be sure continued to be sold to the wider public, but within the community of biologists cracks in the Darwinian foundation began to appear, and these cracks continue to widen today. The next chapter will provide a summary and assessment of several proposed alternatives to Darwinian orthodoxy and, more importantly, the responses to these alternative proposals coming from the biological establishment. It is in the responses that we truly see Darwinian ideology hard at work.

---

101. Woese, "On the Evolution of the Genetic Code," 1550.

7

Challenging the Modern Synthesis

IN HIS PROVOCATIVELY TITLED 1987 article "The Evolutionary Dys-Synthesis," Duke University botanist Janis Antonovics flatly pronounced the failure of the modern synthesis to have solved the riddle of the origin of species. Not only did the modern synthesis fail "in many serious and insidious ways," according to Antonovics, but "at a conceptual level, it may even have hindered rather than furthered our understanding of evolution."[1] Such a pointed critique of the modern synthesis coming from a member of the biological establishment might seem surprising. The grand narrative of Darwinian triumph has proved so successful at overriding such critical voices that most people outside the biological community simply assume that Darwinian evolution is settled science. But nothing could be further from the truth. The biological literature of the last forty years is replete with voices of criticism and proposals for alternatives to or extensions of the modern synthesis. There is a clearly growing recognition among biologists that the modern synthesis simply cannot adequately explain major categories of evidence bearing on the question of evolutionary mechanisms. Much to the chagrin of supporters of the grand narrative, the triumphant nature of Darwinian evolution is proving to be a mere façade that keeps cracking, ironically, under the weight of biological research itself.

In this chapter, I will consider several of these voices of Darwinian critique, organized around three categories of evidence seen by these biologists to pose a challenge to the sufficiency of the Darwinian

---

1. Antonovics, "Evolutionary Dys-synthesis," 321.

mechanism. First is the problem with the fossil record that prompted paleontologists Niles Eldredge and Stephen Jay Gould to pose the theory of punctuated equilibria in 1972. Second I will engage the work of Eva Jablonka and Marion J. Lamb, who are reviving the ghost of Lamarck with their work on epigenetic inheritance, work built upon the directed mutation controversy of the late 1980s and early 1990s. Third, I will consider the problem of the evolution of novelty that has inspired developmental geneticists to consider how whole new classes of organisms have evolved over the history of life, the assumption being that the Darwinian theory of the modern synthesis largely failed in this regard. Finally, this chapter will conclude with an assessment of the place of the modern ID movement within the growing anti-Darwinism of the twenty-first century, an assessment that will avoid the hysterics frequently encountered in Darwinian assessments of ID (as noted in chapter 1). In each case, we will also observe how the grand narrative of Darwinian triumph has tried to ignore, absorb, explain away, or outright ridicule these new trajectories in evolutionary biology; these reactions are further evidence of the grand narrative's ideological nature.

Problems in the Fossil Record

Darwin, as previously noted, was well aware that the fossil record of his day did not support his theory of slow, gradual transformation of organisms over long spans of time. Very few transitional forms were evident in the fossil record linking organisms together into a smooth chain of evolutionary transformation. Darwin hoped that this was due to the deficiency of the fossil record, but time and further discovery has only worsened the situation for Darwin's theory. As late as 1999, Jeffrey Schwartz could say that we are in the dark on the origin of most groups of organisms. "They appear in the fossil record as Athena did from the head of Zeus—full-blown and raring to go, in contradiction to Darwin's depiction of evolution as resulting from gradual accumulation of countless infinitesimally minute variations."[2] Every paleontologist knows that the vast majority of species do in fact appear suddenly in the fossil record, remain virtually unchanged over millions of years, and then just as suddenly disappear into extinction. The fossil record and Darwin's theory are clearly at odds.

---

2. Schwartz, *Sudden Origins*, 3.

Thus in 1972, Niles Eldredge and Stephen Jay Gould tried to address this problem by developing their well-known theory of punctuated equilibria, a theory they viewed as an alternative to Darwinism (the latter being dubbed by them "phyletic gradualism"). According to Eldredge and Gould, for most of the time since Darwin, the majority of paleontologists have been blinded by the emphasis on phyletic gradualism and have therefore interpreted the fossil record through this Darwinian lens, leading to misrepresentations of the process that the fossil record actually attests to:

> We believe that an inadequate picture has been guiding our thoughts on speciation for 100 years. We hold that its influence has been all the more tenacious because paleontologists, in claiming that they see objectively, have not recognized its guiding sway. We contend that a notion developed elsewhere, the theory of allopatric speciation, supplies a more satisfactory picture for the ordering of paleontological data.[3]

What is *allopatric speciation*?

Eldredge and Gould theorize that since the fossil record preserves so few transitional forms, new species must have evolved rapidly in small, peripheral populations cut off from an organism's main population area. That is, due to some sort of environmental disruption, a small portion of a population becomes geographically cut off from the main population and remains this way for an extended period of time, which prohibits gene flow between the peripheral population and the main population. Rapid evolution then occurs in the peripheral population, leading to the development of a new species distinct from the main population. This would explain why the origin of new species is so hard to identify in the fossil record. Eldredge and Gould write, "A consequence of the allopatric theory is that fossil species do not originate in the place where their ancestors lived. This is why we cannot trace the gradual splitting of species in the fossil record and why we find so few transitional forms."[4] Under the mantra "stasis is data," Eldredge and Gould take seriously the fossil record's unquestioned attestation to the fact that once they form, most species appear not to change appreciably over the span of their existence. Even a "living fossil" like the horseshoe crab has barely changed over one hundred million years of evolutionary history. The lack of evidence

---

3. Eldredge and Gould, "Punctuated Equlibria," 86.
4. Ibid., 94.

for continuous, gradual evolutionary change cannot be dismissed, but it must be understood as evidence for an alternative understanding of evolutionary mechanisms.

Eldredge and Gould readily admit that the theory of allopatric speciation (also referred to as *geographic speciation*) did not originate with them.

> Most paleontologists, of course, are aware of this theory, but the influence of phyletic gradualism remains so strong that discussions of geographic speciation are almost always cast in its light: geographic speciation is seen as the slow and steady transformation of two separated lineages—i.e., as *two* cases of phyletic gradualism.[5]

As we will soon see, the ubiquitous Ernst Mayr will take credit for developing the theory of allopatric speciation. What Eldredge and Gould are arguing is that Darwinians (presumably including Mayr) were interpreting allopatric speciation through the lens of their commitment to phyletic gradualism, thereby missing the true importance of this theory as an alternative to phyletic gradualism. Eldredge and Gould were exhorting their fellow paleontologists to embrace the more radical picture of evolution that emerges from an engagement with the fossil record unencumbered by a prior commitment to Darwinism. In their view, "the norm for a species or, by extension, a community is stability. Speciation is a rare and difficult event that punctuates a system in homeostatic equilibrium."[6] Seeing stability rather than continuous change as the norm is a clear challenge to phyletic gradualism.

Five years after their initial article on punctuated equilibria, Gould and Eldredge (the order of the names had switched) continued to voice their concerns with phyletic gradualism, calling it a restrictive dogma that excludes a priori the very data that might refute it. "Stasis is ignored as 'no data,' while breaks are treated as imperfect data."[7] Gould and Eldredge argue that stasis *is* important data for how evolution unfolds: not by smooth gradual transitions but by fits and starts. They observe that at higher levels of the taxonomic hierarchy, phyletic gradualism has always been in serious trouble even if it does remain the "official" position of most Western evolutionary biologists. Smooth transitions

---

5. Ibid. (Italics original).
6. Ibid., 115.
7. Gould and Eldredge, "Punctuated Equilibria," 119.

between body plans "are almost impossible to construct, even in thought experiments."[8] This is the reason, according to Gould and Eldredge, that a committed Darwinian like George Gaylord Simpson had to invoke the concept of quantum evolution. By 1980, Gould would still be able to say: "I have been reluctant to admit it . . . but if Mayr's characterization of the synthetic theory is accurate, then that theory, as a general proposition, is effectively dead, despite its persistence as textbook orthodoxy."[9] He then invokes the work of the much vilified Richard Goldschmidt to answer the question of how new species arise so rapidly in peripherally isolated populations, and concludes, "I envisage a potential saltational origin for the essential features of key adaptations."[10] Over a period of eight years, Gould and Eldredge clearly viewed their theory of punctuated equilibria as a major challenge to the Darwinian evolutionary mechanism. How did the Darwinians respond?

Given Eldredge and Gould's status as respected paleontologists and the fact that the fossil record is clearly more consistent with punctuated equilibria than it is with phyletic gradualism, Darwinian advocates could not simply ignore or dismiss the theory of punctuated equilibria. Instead, they simply absorbed the new theory into the Darwinian framework. But not necessarily in a convincing way. Douglas Futuyma, for example, argued that nothing in traditional evolutionary theory was violated by Eldredge and Gould's theory except for one (what he labeled) heretical point: the idea that species get genetically locked in and do not evolve for long periods of time—the concept of stasis. In Futuyma's words, "Except for this admittedly important point, the difference between punctuated equilibrium and the orthodox view is virtually nonexistent."[11] But this is like me saying that despite the fact that I am only five feet seven inches tall, the difference between me and Lebron James in our ability to slam-dunk a basketball is virtually nonexistent! Futuyma can absorb punctuated equilibria into a Darwinian framework only by completely downplaying the significance of the evidence that renders punctuated equilibria a fundamentally different (and, according to Futuyma, heretical) theory.

---

8. Ibid., 147.
9. Gould, "Is a New and General Theory," 120.
10. Ibid., 127.
11. Futuyma, "Evolution as Fact and Theory," 4.

For his part, Ernst Mayr admitted to the great influence punctuated equilibria had on paleontology and evolutionary biology. But he believed the term *punctuated equilibria* was more novel than the actual concept it named. A role for peripheral populations had been around, according to Mayr, since 1825, though this role was minimized after the population geneticists emphasized that evolution proceeded most rapidly in populous, widespread species. Mayr himself worked to recover the importance of allopatric speciation and claims that Eldredge and Gould based their theory on his, though they placed a much stronger emphasis on the importance of evolutionary stasis. Mayr admits that a modest theory of punctuational evolution is strongly supported by the facts, but he also believes that it fits fully into a Darwinian framework—so much so that Mayr expressed surprise that Eldredge and Gould's theory provoked hostility.[12] Mayr, not surprisingly, does go on to chide Gould for invoking Goldschmidt to explain how new species evolve rapidly in peripherally isolated populations, and claims that Gould misrepresents the heretical nature of Goldschmidt's views. For Mayr:

> The argument, thus, is not whether phenotypic saltations are possible, but rather whether evolution advances through the production of individuals representing new types or through the rapid transformation of populations. No matter how rapid, such a populational "saltation" is nevertheless Darwinian gradualism.[13]

Just like Futuyma, Mayr absorbs punctuated equilibria into the framework of phyletic gradualism, essentially misrepresenting the truly non-Darwinian character of the theory as Eldredge and Gould originally envisioned it.

The controversy over the meaning of the fossil record for evolutionary mechanisms has not abated, as we saw earlier in our discussion of the Cambrian explosion. In 2000, the paleontologist Robert Carroll would argue the need for a new evolutionary synthesis with the words: "Increasing knowledge of the fossil record and the capacity for accurate geological dating demonstrate that large-scale patterns and rates of evolution are not comparable with those hypothesized by Darwin on the basis of extrapolation from modern populations and species."[14] If one takes the

---

12. Mayr, *Toward a New Philosophy of Biology*, 462.
13. Ibid., 468.
14. Carroll, "Towards a New Evolutionary Synthesis," 27.

fossil record at face value, the seeming sudden emergence of new species makes the process of evolution appear far more mysterious than Darwin envisioned. No wonder the grand narrative of Darwinian triumph has tried to override the mystery by tendentiously forcing a Darwinian framework over the fossil record. The fossil record was an embarrassment to Darwin himself and continues to be today for the advocates of his theory.

Epigenetic Inheritance

In 1995, Israeli biologist Eva Jablonka and her British collaborator Marion J. Lamb revived the ghost of Lamarck with the publication of *Epigenetic Inheritance and Evolution: The Lamarckian Dimension*. In the grand narrative of Darwinian triumph, the Lamarckian principle of the inheritance of acquired characters was supposed to have been rendered obsolete in the early twentieth century by August Weismann's famous experiment of cutting off the tails of mice to show that succeeding generations do not display either missing or shorter tails. Weismann had pronounced the principle of the isolation of germs cells from somatic cells in an organism—meaning that changes occurring in the somatic cells during the lifetime of an organism cannot be transmitted to the next generation since traits are inherited only through the isolated germ cells. This principle became a bedrock of Darwinian evolution. It established that the variation on which natural selection acts can only arise from random changes occurring in DNA base sequences during cell replication.

But Jablonka and Lamb challenged this sacred article of Darwinian orthodoxy by considering the way that epigenetic inheritance might make a subtle form of inheritance of acquired characters—a modified Lamarckism—possible. And they knew they were treading on dangerous ground when advancing this theory, for they begin their book with the surprising statement: "This book was written because we believe that it is time to re-examine the role of the inheritance of acquired characters in evolution. We are aware that this statement is likely to produce an immediate negative reaction in the reader."[15] They then continue:

> We started writing this book as a response to the mixture of enthusiasm and antagonism that we found for the idea that epigenetic inheritance is important in evolution. The emotional

---

15. Jablonka and Lamb, *Epigenetic Inheritance and Evolution*, vii.

and aggressive reactions that we often encountered when presenting our views made it clear that we were touching sensitive nerves.[16]

One would expect Jablonka and Lamb to be able to present their scientific argument and have it be assessed based on its evidentiary strength without their feeling the need at the outset to acknowledge the "controversial" nature of what they are proposing and to note the "emotional and aggressive reactions" they had already experienced when sharing their ideas. This is strong evidence for the ideological nature of the Darwinian narrative that Jablonka and Lamb knew they were challenging.

So what is *epigenetic inheritance*, and why is it deemed so controversial? If we consider a human body, it is well known that all the cells making up that body turn over about every seven years. Old cells die and are replaced by new ones. Since all the cells in a human body have identical DNA in their nuclei but not all cells function in the same way (some are bone cells, some are muscle cells, and so forth), how genes are expressed determines how a cell comes to function. A bone cell and a muscle cell may have identical DNA, but the two cells express their genes differently, leading to their different functions. When an old bone cell dies to be replaced by a new one, there must be a mechanism whereby the old cell can transmit its unique pattern of gene expression to the new cell so that the new cell can pick up where the old cell left off and take on its function in the body. This is known as cell memory, and cell memory is transmitted via an epigenetic process rather than a purely genetic one (since there are no changes in the DNA sequences between the two cells).

The technicalities of how information is transmitted from one cell to another epigenetically is beyond our scope here. But good evidence points to the likelihood that changes occurring in the somatic cells of an organism as it interacts with its environment can produce what are called epigenetic tags on DNA (like methylation)—tags that influence how genes are expressed. These epigenetic tags can in some cases be passed on to the next generation, raising the possibility that acquired characters are indeed heritable, meaning that the environment can directly drive evolutionary change. Jablonka and Lamb argued that some new inherited variations do in fact arise as a direct, and sometimes even a directed, response to environmental challenge.[17] Any talk, of course, of evolu-

16. Ibid., x.
17. Ibid., 1.

tion being a directed response to environmental challenge will raise the specter of teleology, and Darwinians will balk. Hence, the "emotional and aggressive reaction" to Jablonka and Lamb's work.

Jablonka and Lamb fully recognize what is at stake in their theory:

> Perhaps the foremost reason for the reluctance to accept Lamarckian interpretations is the feeling that by so doing, one is accepting purposeful evolutionary responses: that an organism has some indefinable properties that propel it towards some goal. How does the organism *know* how to change its genetic material according to new environmental specifications?[18]

Not wanting to stray too far off the Darwinian reservation, Jablonka and Lamb decided to bow to the dictates of the grand narrative and to assure their readers of their Darwinian bona fides. They point to the debate on directed mutations that we saw in the previous chapter as providing a way to circumvent any teleological implications. They cite studies by Bruce Wallace and Sydney Brenner, who both "attempted to make the Lamarckian idea of environmentally directed mutation more palatable to Darwinians by treating it as an adaptive response, which has itself evolved through Darwinian selection of random variations."[19] Jablonka and Lamb then state unequivocally, "We do not doubt that the basic mechanisms underlying the inheritance of acquired variations evolved in a Darwinian fashion by the selection of accidental variations."[20] So they assure us that a non-Darwinian mechanism of evolution (epigenetics) was still produced by a Darwinian one (natural selection), allowing Darwinians to sleep at night, but at the expense of undercutting the revolutionary nature of their own work. Why?

On this question Jablonka and Lamb are most forthright: "We feel it necessary to stress our belief in Darwinian evolution because recent history has shown that any argument suggesting that Darwinian evolutionary theory should be modified or amended is liable to be used by creationists as evidence that the theory of evolution is wrong."[21] To be taken seriously as biologists, Jablonka and Lamb simply cannot allow themselves to be understood as providing comfort to the creationist enemy, so they relent. But their earlier acknowledgement that they know their

18. Ibid., 24.
19. Ibid.
20. Ibid., 25.
21. Ibid., 27 n. 1.

book will produce an immediate negative reaction in the reader is strong evidence that Jablonka and Lamb's readers well know that epigenetic inheritance does pose a fundamental challenge to Darwinian evolution, no matter how hard Jablonka and Lamb try to calm their readers' nerves.

How was Jablonka and Lamb's book received? Their work could not be ignored since the existence of epigenetic inheritance is well documented. Its importance as an evolutionary mechanism is what is at issue. In a review, Graham Bell admitted that strange exceptions to general rules can occasionally occur, but without greatly disturbing our overall worldview: "So long as they are rare exceptions, they remain an interesting footnote to a selectionist text, and despite the skill with which Jablonka and Lamb have assembled their arguments and their material, I think this is the most likely outcome."[22] In other words, no need to worry. Jablonka and Lamb have simply demonstrated some "rare exceptions" to the Darwinian rule. But the rule still stands!

Koen Martens sharply disagrees. He charges evolutionary geneticists with making a priori assumptions about the random nature of variation—assumptions they adhere to "with nearly religious zeal."[23] Jablonka and Lamb's book, he believes, fundamentally challenges this dogma. In his view, integrating their work into the modern synthesis would require a substantial change to Darwinian theory, and he views Jablonka and Lamb's book as having "all the markings of a classic."[24]

Martens was not alone in his enthusiasm. Mae-wan Ho went even further, calling the Weismannist assumption that genes are insulated from environmental influence "flawed from the start," while highlighting the evidence for fluid genome processes like transposable elements, directed mutation, and immune system processes.[25] According to Ho, Darwinian evolution was almost fatally undermined by the discovery of these fluid genome processes, and trying to integrate them into a Darwinian framework is misguided at best: "To insist on calling those instances of clearly directed genetic changes "selection" is to engage a fine point of semantics in order to save an hypothesis so riddled with holes that one wonders why it is not allowed to sink of its own accord."[26]

22. Bell, Review of *Epigenetic Inheritance and Evolution*, 267.
23. Martens, Review of *Epigenetic Inheritance and Evolution*, 341.
24. Ibid., 342.
25. Ho, "Why Lamarck Won't Go Away," 82.
26. Ibid., 83.

Whereas Ho clearly agrees with Jablonka and Lamb, she does not feel the need to bow to Darwinian orthodoxy. Biology, she argues, needs to move beyond the "mechanical era" like the rest of science. Biology must begin to focus on the organism as an integrated whole that is "open to the environment, to which it responds as much as it transforms."[27]

For their part, Jablonka and Lamb continued to press their modified Lamarckism, as they also continued to try and fit their controversial views into a broader Darwinian framework. In a 2005 publication, Jablonka and Lamb offered additional evidence for the directed nature of at least some variations and the inheritance of acquired characters, which they realized continued to sound like a "grave heresy" to many biologists.[28] DNA works, they argue, because organisms have mechanisms that can monitor DNA replication and repair errors. But how did this sophisticated ability develop?

> This amazing system for maintaining the integrity of DNA has *presumably* evolved through natural selection for DNA-caretaker genes. Lineages with poor DNA maintenance and sloppy replication failed to survive because they kept changing, producing all sorts of new mutations, most of which were detrimental.[29]

Once again, Jablonka and Lamb feel the need to explain the development of a non-Darwinian process by appeal to a Darwinian one, even though they acknowledge by use of the word "presumably" that they have no real evidence that natural selection did or could in fact evolve DNA-caretaker genes.

In the end, Jablonka and Lamb were so strongly influenced by the larger ideological battles animating modern biology that they felt it necessary to constantly monitor their language and explicitly deny the more radical implications of their own work. When commenting on the phenomenon of directed mutation, an idea to which they fully ascribe, we are told: "'Directed mutation' is part of the jargon of genetics, and it does not mean that we or other biologists believe there is some guiding intelligence or 'hand of God' directing changes in DNA according to the organism's needs. Such ideas have no part in scientific reasoning (and through their absurdity ridicule religion as well)."[30] Once again, Jablonka

27. Ibid., 84.
28. Jablonka and Lamb, *Evolution in Four Dimensions*, 7.
29. Ibid., 86 (italics added).
30. Ibid., 87.

and Lamb argue that natural selection produced mechanisms that alter DNA in response to cellular and environmental signals. Everything must be traced back to the omnipotent, creative force of natural selection! Curiously, Jablonka and Lamb note that at least 45 percent of the human genome is composed of transposable elements, and as much as 50 percent in plants. Transposable elements, therefore, must have played a role in evolution. "Mobilizing them when times are bad could undoubtedly produce a lot of new genetic variation."[31] Perhaps. But what is responsible for "mobilizing" them? This sounds suspiciously teleological, Jablonka and Lamb's protests notwithstanding.

But perhaps their protests are understandable. In a review of this later work, Jonathan Grose comments:

> Chapter 9 offers plausible accounts of the origin of each inheritance system to refute a teleological or purposive view of evolution and thereby make a straightforward denial of so-called ID theory. Such a rebuttal is probably unnecessary for many readers but does serve to forestall the highjacking of controversy over details of Darwinian evolution by proponents of intelligent design.[32]

Fear of emboldening creationists and ID advocates is once again explicitly stated as a major reason to smooth over controversies in evolutionary theory. Darwinian natural selection becomes an ideological bulwark against the "nonscientific" hoards ramming against the gates of the biological establishment. But while Darwinism continues to play a useful ideological role in the protection of the guild interests of the biological establishment, its ideological function remains irrelevant to its truth-value.

The case for epigenetic inheritance as a challenge to Darwinism has recently been given full voice by eminent Oxford physiologist Denis Noble in his 2017 book *Dance to the Tune of Life: Biological Relativity*. Taking his cues from Einstein's famous theory of relativity, Noble argues that in biological systems, no level of causation from molecules up to entire organisms can be privileged over any other. Standing against the genecentric view of modern molecular biology on which Darwinian evolution is based, Noble notes that predicting disease from family history and lifestyle is far more reliable than predicting it from genome sequencing. Knowing the DNA sequence of an organism provides little in the way of

---

31. Ibid., 250.
32. Grose, Review of *Evolution in Four Dimensions*, 670.

predictive power about the form that an organism's phenotype will take. Gene expression is what matters, but it is the cells, tissues, and organs of the body that control gene expression. Therefore, identical genotypes can produce vastly different phenotypes if the cells, tissues, and organs of two different organisms control gene expression differently. Based on this, Noble pronounces the principle of biological relativity: "All levels in organisms have causal efficacy. There is no privileged level from which all others may be derived."[33]

For Noble, this principle of biological relativity has enormous implications for how we think of evolutionary mechanisms. He states, "The experimental evidence that wholesale reorganization of genomes has occurred during evolution shows that genomes did not always evolve by gradual accumulation of random small mutations."[34] He points to the work of the developmental biologist of the 1950s Conrad Waddington, who provided evidence for the inheritance of acquired characters. Developmental studies of fruit flies demonstrated that embryos could be persuaded to develop different wing structures simply by changing the environmental temperature or through a chemical stimulus, without any change in the underlying DNA sequence. And these changes could be inherited. About Waddington, Noble writes:

> His text could not be clearer. Yet his work was sidelined. I find it shameful that this kind of denigration can be done in the name of science, just as it is shameful that Lamarck is not recognized for what he achieved. These are the reasons why the dogmatism of Neo-Darwinist popularization has been damaging. What was missed was the opportunity to develop an integrated, and what I would call a relativistic, theory of evolution over 50 years ago. This is a long time for an important lead to be ignored.[35]

Jablonka and Lamb may have ultimately assented to Darwinian pressure, but not Noble. He makes no attempt to prove his Darwinian bona fides. For him, Darwinism is a dead end. Why has the evidence undermining Darwinism been ignored for so long? Noble relates a conversation he had about biological relativity with an eminent neuroscientist, who responded, "I would go along with you, Denis, except that what you

---

33. Noble, *Dance to the Tune of Life*, 181.
34. Ibid., 201.
35. Ibid., 218.

are arguing would let God back in." Reflecting on this statement, Noble continues:

> This was very revealing to me, since I certainly did not think that conclusion followed at all. Yet he clearly did, and had no doubt about it. In fact, I suspect that for many scientists, defending reductionism, including particularly Neo-Darwinism, was a necessity in order to counter the claims of creationist religion or supernatural intelligent design. I think they are throwing out the baby with the bathwater.[36]

If there is evidence that evolution can be driven by processes going on at the level of cells, tissues, organs, and even whole organisms, such evidence must be included in any comprehensive scientific understanding of evolution. Noble seems keenly aware, however, that its dismissal is driven by ideological, not scientific concerns. Noble's work provides just one more example of how the grand narrative of Darwinian triumph continues to override the search for truth, and potentially subverts religion in the process.

The Evolution of Novelty

Some contemporary evolutionary biologists may be troubled by the recalcitrance of the fossil record to support Darwinian evolution, and by the stubborn refusal of Lamarckism to remain dead and buried. Others are unnerved by the question of the evolution of novel structures. Natural selection can explain the fate of new structures once they exist, but can it explain the development of whole new kinds of organisms and body plans? In the words of Swarthmore College biologist Scott Gilbert, "The modern synthesis is remarkably good at modelling the survival of the fittest, but not good at modelling the arrival of the fittest."[37] Gerd Müller and Stuart Newman agree, calling the failure to incorporate innovation "one of the major gaps in the canonical theory of evolution."[38] I will consider in this section several approaches to this question.

---

36. Ibid., 248.
37. Cited in Whitfield, "Postmodern Evolution?" 282
38. Müller and Newman, eds., *Origination of Organismal Form*, 4.

Evolutionary Emergence

In 1985, Robert G. B. Reid published *Evolutionary Theory: The Unfinished Synthesis* to signal his view that the modern synthesis remained incomplete because it could not explain the origin of organismal form. He followed this up in 2007 with *Biological Emergences: Evolution by Natural Experiment* to further explicate the insufficiencies of the modern synthesis as he understood them. I will briefly engage this later work.

Reid's writing is notoriously complex and resistant to easy summarizing. But if he is clear on one point, it is that his concept of "natural experiment" as the driving force of evolution is meant to replace the term *natural selection*, not to be an addendum to it. Reid writes: "As an alternative to the Modern Synthesis, I suggest that we work toward an Emergence Theory that accounts for generative mechanisms and the emergent qualities of their evolutionary products."[39] Reid believes there has been an overemphasis on natural selection in evolutionary thinking and complains that "every generation of undergraduates is brainwashed into this way of thinking—'If it exists, it must have been selected.' Biology is *constantly* being misrepresented to students and the public at large."[40] In fact, Reid argues that in some cases natural selection might act as an impediment to evolutionary change by adapting an organism so well to its environment that the emergence of a novel structure gets selected out of the population.

So what does Reid mean by "natural experiment"? He draws on a variety of controversial processes like directed mutation, epigenetic inheritance, and developmental genetics to argue that evolution proceeds in a saltatory way, as organisms generate novel structures and then test them in the crucible of adaptation and survival. Some experiments work and are preserved, some don't. But in Reid's view, novel structures cannot come into existence through the slow, gradual selection of minute variations, as Darwin's theory requires. There must be processes at work allowing for the rather sudden emergence of novel structures. Reid admits that his theory shares some commonality with Eldredge and Gould's punctuated equilibria but criticizes the latter for backing down in the face of pressure from the Darwinian establishment and opting for

---

39. Reid, *Biological Emergences*, 21.
40. Ibid., 43.

"respectability over revolution."[41] Reid is clear that he is advocating for a revolution in evolutionary thought.

To emphasize his radical departure from Darwinism, Reid chastises Darwinian biologists for dismissing the concept of evolutionary progress as a delusion. Any talk of progress sounds too teleological, so Darwinians tend to shy away from seeing any kind of direction or progress in the history of life. In response Reid quips, "But here *we* are, discussing it! And *there* are our bacterial ancestors, who aren't!"[42] For Reid, it is obvious that evolution has progressed over time in the direction from simplicity towards greater complexity (which Lamarck had argued), and Darwinians ignore this evident truth for ideological reasons. There must be a generative process underlying this directionality. A process based simply on the randomness of variation could not produce it. Finally, Reid cites a famous quote by J. B. S. Haldane: "The world is not only queerer than anyone has imagined, but queerer than anyone can imagine." From this Reid asks, "Is it not time that we confronted that queerness and tried to understand it instead of fleeing from it?"[43] Reviewers of Reid's work have differed sharply over how to answer this question.

Not surprisingly, Ernst Mayr negatively reviewed Reid's 1985 book that labeled the synthesis "unfinished." Also not surprisingly, he criticized Reid for being a typologist who is unable to engage in population thinking. (This is Mayr's repeated criticism of anyone who rejects natural selection, as we have already seen.) Mayr concedes that the variation side of the variation-selection mechanism is still insufficiently understood, but he does not believe Reid has shed any helpful light on the problem: "Uninformed attacks on natural selection cannot make a constructive contribution."[44]

But reviewing the same book, John F. Cornell came to a quite different conclusion. He points out that most historians of evolutionary theory have taken a defensive stance toward Darwinism, while those few who have taken a more critical stance (like Getrude Himmelfarb) could be dismissed as nonscientists who do not fully grasp the scientific nuances of the theory. But, reviewer Cornell continues, "Robert G. B. Reid is a practicing biologist in several specialized fields, and his readers

---

41. Ibid., 403.
42. Ibid., 364 (italics original).
43. Ibid., 431.
44. Mayr, Review of *Evolutionary Theory*, 359.

will notice that scientifically he is quite well informed. He is one of the more thoughtful recent students of evolutionary theories."[45] For Cornell, Reid took seriously the vice that attended Darwin's great virtue as a scientist: His "eager subordination of all evolutionary problems to the problem whose solution he happened to understand, that of ecological adaptation."[46] Cornell was clearly more impressed than Mayr with Reid's provocative work.

In response to Reid's 2007 book on natural experiments Christopher Rose was even more effusive:

> Reid argues convincingly that the selectionist paradigm is a conceptual dead-end for understanding innovation since it mistakenly views natural selection as a creative force in evolution. Natural selection is a metaphor for processes (differential survival and reproduction) that happen after events that generate innovation (the natural experiments upon which Reid builds his emergence paradigm)."[47]

Rose understands Reid's championing of biological emergence as a step in the right direction. Evolutionary biology needs to be pushed beyond a reductionist and gene-centered explanation for properties like multicellularity, the evolution of novel body plans, behavioral flexibility, self-maintenance, and the evolution of human intelligence. But Rose also acknowledges a significant downside to Reid's theory: "Emergentism will undoubtedly increase the challenge of teaching evolution theory and convincing the public (and ourselves) that biologists know what we are talking about."[48] The role played by guild interests in maintaining the grand narrative of Darwinian triumph could not be better stated. What biologist will want to admit to students and to the public that evolutionary mechanisms truly remain a mystery a century and a half after Darwin?!

The Altenberg 16

In July of 2008, sixteen evolutionary biologists convened at the Konrad Lorenz Institute for Evolution and Cognition Research in Altenberg, Austria, to discuss the modernizing or extending of the modern

---

45. Cornell, Review of *Evolutionary Theory*, 424.
46. Ibid.
47. Rose, Review of *Biological Emergences*, 871.
48. Ibid., 873.

synthesis. Dubbed the Altenberg 16, this conference was reported on in the scientific journals *Science* and *Nature* and became the subject of a book by science reporter Suzan Mazur, *The Altenberg 16: An Exposé of the Evolution Industry*. The participants in this conference, some of whom we have already met in the pages of this book—Massimo Pigliucci, Stuart Newman, Eva Jablonka—feel that the modern synthesis is not sufficient to account for the development of organismal complexity and want to extend the synthesis in a variety of ways, with the inclusion of developmental biology being one of the most important. Development (the process by which a fertilized egg becomes a full-blown organism) was largely ignored in the modern synthesis. And some Darwinians even today prefer to continue ignoring it.

In her reporting on the conference, Elizabeth Pennisi quotes ardent Darwinian Jerry Coyne as saying, "It's a joke. I don't think there is anything that needs fixing."[49] In his reporting in *Nature*, John Whitfield quotes Coyne as commenting, "The whole thing about natural selection being an insufficient paradigm seems grossly overblown."[50] Why does he think this? Coyne continues, "People shouldn't suppress their differences to placate creationists, but to suggest that neo-Darwinism has reached some kind of crisis point plays into creationists' hands."[51] Once again, the specter of creationists crouching at the door ready to strike becomes a rationale for holding to the Darwinian paradigm.

According to Stuart Newman, who researches principles of self-organization in biological organisms, self-organization does represent a challenge to Darwinian evolution. But he continues, "People are concerned—though I don't agree with them for being concerned about it—but people are concerned that if they open up the door to non-Darwinian mechanisms, then they're going to let the creationists slip through the door as well."[52] University of Arizona cognitive scientist Massimo Piatelli-Palmarini is even more direct:

> Many biologists are reluctant to steer away from natural selection. If they accept non-selectionist mechanisms as important in evolution, they still argue that the Darwinian paradigm is only modified, not subverted ... They fear that the tenants of

---

49. Pennisi, "Modernizing the Modern Synthesis," 196.
50. Whitfield, "Postmodern Evolution?" 283.
51. Ibid., 284.
52. Cited in Mazur, *Altenberg 16*, 132.

intelligent design and the creationists (people I hate as much as they do) will rejoice and quote them as being on their side. They really fear that, so they are prudent, some in good faith, some for calculated fear of being cast out of the scientific community.[53]

If a biologist can be cast out of the scientific community or otherwise marginalized (by having research funding removed or tenure denied)—or even if there is a palpable fear of this—simply for voicing substantive scientific criticisms of Darwinian evolution, then we know that the grand narrative of Darwinian triumph is being enforced by disciplinary measures, and this is clear evidence of this narrative's ideological nature and function.

Despite the dangers of dissenting from Darwinism, the work of the Altenberg 16 was published in a 2010 volume edited by Massimo Pigliucci and Gerd Müller bearing the title *Evolution: The Extended Synthesis*. In their introduction, Pigliucci and Müller write:

> Several approaches discussed in this volume show that nongradual change is a property of complex dynamical systems, including biological organisms, and that various kinds of mechanisms for discontinuous change are now known from the domains of genome evolution, phenotypic plasticity, epigenetic development, and nongenetic inheritance.[54]

They further argue that the discovery of newer molecular processes have clearly demolished the alleged "central dogma" of molecular biology, the term *dogma* being a word that "arguably should never be used in science."[55] If they are right that the Central Dogma has been demolished (scientists like Barbara McClintock and James Shapiro would agree that it has), then there seems to be no reason other than ideology to continue to enforce Darwinian orthodoxy. Evolution is a far more complex and mysterious process than Darwin or the shapers of the modern synthesis could have ever imagined.

Facilitated Variation

Two members of the Altenberg 16, Marc W. Kirschner and John C. Gerhart, have developed a theory to explain the origin of novelty called

53. Ibid., 317.
54. Pigliucci and Müller, eds., *Evolution*, 13.
55. Ibid., 9.

facilitated variation. While this may sound like an idea based on the concept of directed mutation, Kirschner and Gerhart actually pronounce the failure of the famous experiment performed by John Cairns and his colleagues while also pronouncing the failure of all attempts to demonstrate Lamarckian inheritance (though they ignore the work of Eva Jablonka and Marion Lamb). Rather, facilitated variation is based on the notion of conserved, core processes and components that allow an organism to respond to environmental challenge by drawing on what is already at work in the phenotype of the organism. "Novelty mostly draws on what is already present in the phenotype."[56] In other words, these conserved, core processes allow an organism, through small changes to its regulatory systems, to undergo significant phenotypic change without having to build up new structures in a gradual, Darwinian way. According to Kirschner and Gerhart, an organism responds to mutation "by making changes it is largely prepared in advance to make."[57] That is, random mutation does not create phenotypic variations but rather channels those variations through regulatory changes. "Rather than staggering like a drunken sailor, evolution marches along a myriad of paved pathways."[58] The technicalities of what constitutes a conserved, core process need not concern us here (though the genetic code is one of them). The important point is that according to Kirschner and Gerhart, these processes developed very early in the history of life and have been conserved through evolutionary history.

Does facilitated variation really solve the problem of the evolution of novelty? Not necessarily. Kirschner and Gerhart state, "Although we are left with a more complete theory of evolutionary change, we still have the question of how facilitated variation arose in the first place and why it has been maintained."[59] If they cannot account for the origin of the conserved, core processes that make facilitated variation possible, then they really haven't solved the problem. Kirschner and Gerhart admit that "the most obscure origination of a core process is the creation of the first prokaryotic cell. The novelty and complexity of the cell is so far beyond anything inanimate in the world of today that we are baffled by how it

---

56. Kirschner and Gerhart, *Plausibility of Life*, 224.
57. Ibid., 226.
58. Ibid., 247.
59. Ibid., 248.

was achieved."[60] In another place they write, "Everything about evolution before bacteria-like life forms is sheer conjecture."[61] Given these admissions, it is not surprising to learn that Kirschner and Gerhart's theory received an underwhelming response from their biologist colleagues.

Robert Perlman calls Kirschner and Gerhart's claim to have produced a major, new scientific understanding of evolution an exaggeration. The theory is simply a useful synthesis and extension of the ideas of others.[62] In the view of Brian Charlesworth, "Kirschner and Gerhart do not present any detailed examples of how the properties of developmental systems have actually contributed to the evolution of a major evolutionary novelty."[63] And in the opinion of Carl D. Schlichting, "The perspective of facilitated variation is unlikely to capture the fancy of anyone who is content with the notion that organismal diversity is the product of the timeless sculpting of natural selection on the raw materials generated by random mutation."[64] Although these reviewers are not convinced by Kirschner and Gerhart's theory, they avoid pointing out its major flaw: its failure to account for the evolution of the core processes themselves. This is likely because they know that Darwinian evolution also fails to explain the origin of life from nonlife or the origin of the genetic code or the origin of the first prokaryotic (bacterial) cell. Darwinism and facilitated variation share some of the same fatal flaws. So the reviewers pass.

This may be understandable, but far more significant is how they let pass Kirschner and Gerhart's most controversial statement:

> The road to evolutionary change is paved with physiological adaptability. Phenotypic variation, and along with it evolutionary change, is facilitated by simple regulatory tweaks to existing physiological and developmental processes that long ago were *designed* so that the organism could adapt to its environment.[65]

Unable to account for the origin of conserved, core processes as part of a random, unguided process, Kirschner and Gerhart unexpectedly insert the dreaded teleological word *design* into their explanation. It is not hard to see why. They admit to being baffled by how the novelty and complexity

60. Ibid., 256.
61. Ibid., 50.
62. Perlman, Review of *The Plausibility of Life*, 314.
63. Charlesworth, "On the Origins of Novelty and Variation," 1620.
64. Schlichting, Review of *The Plausibility of Life*, 170.
65. Kirschner and Gerhart, *Plausibility of Life*, 37 (italics added).

of the cell was achieved. But maybe this bafflement would be relieved if we just considered the possibility that life does show signs of being designed. This is not a new idea, but formed the core of William Paley's seminal 1802 work on natural theology and its famous watchmaker analogy. Today, this is the idea animating the best-organized anti-Darwinian movement in the world, the modern ID movement. No survey of non-Darwinian evolutionary theories can ignore this growing phenomenon.

Intelligent Design

As we saw in chapter 1, the biological establishment treats the modern ID movement with contempt. Rather than engage their ideas, biologists denigrate ID advocates as bad scientists motivated primarily by religious faith, branding the movement "Intelligent Design Creationism," a title ID advocates never use for themselves. This branding allows the biological establishment to lump ID advocates in with young earth creationists, who take their cues from a literal reading of the biblical book of Genesis and deny evolution altogether. Since strict creationism is relatively easy to dismiss on evidentiary grounds, lumping ID in with it allows the biological establishment to dismiss ID simply by association without having to substantively engage its specific arguments. ID, however, is not creationism. It is a far more sophisticated theory that its adherents are convinced is just as scientific as Darwinian evolution.

Of course, the status of ID as science is highly contested. Because its adherents are seen as being motivated primarily by conservative Protestant religion, the argument goes that ID cannot, by definition, be science. Many ID advocates do indeed come from religious backgrounds, but their religious commitments are far more diverse than are usually portrayed. Among influential figures in the movement we find a conservative Roman Catholic, several evangelical Protestants, a member of the Unification Church, an Orthodox Jew, a secular Jew, and even a secular Dane who dabbles in Hinduism. Moreover, claiming that ID proponents are motivated by religion is not really saying anything of substance. Many high-profile proponents of Darwinian evolution have clear naturalistic ideological commitments that likely influence their interpretation of scientific data as much as is the case with ID researchers. It will be helpful to give some examples to support this crucial point.

The noted atheist Richard Dawkins has famously said, "Although atheism might have been *logically* tenable before Darwin, Darwin made it possible to be an intellectually fulfilled atheist."[66] He then elaborated on this point several years later:

> In a universe of physical forces and genetic replication, some people are going to get hurt, other people are going to get lucky, and you won't find any rhyme or reason in it, nor any justice. The universe that we observe has precisely the properties we should expect if there is, at bottom, no design, no purpose, no evil and no good, nothing but blind, pitiless indifference.[67]

What does this mean for the nature of human life? Douglas Futuyma responds, "Some shrink from the conclusion that the human species was not designed, has no purpose, and is the product of mere mechanical mechanisms—but this seems to be the message of evolution."[68] What, then, does this mean for morality and ethics? William Provine has a ready answer:

> Modern science directly implies that there are no inherent moral or ethical laws, no absolute guiding principles for human society . . . We must conclude that when we die, we die, and that is the end of us . . . Finally, free will as it is traditionally conceived—the freedom to make uncoerced and unpredictable choices among alternative courses of action—simply does not exist . . . There is no way that the evolutionary process as currently conceived can produce a being that is truly free to make moral choices.[69]

Finally, E. O. Wilson fully tips his ideological hand:

> We have come to the crucial stage in the history of biology when religion itself is subject to the explanations of the natural sciences. As I have tried to show, sociobiology can account for the very origin of mythology by the principle of natural selection acting on the genetically evolving material structure of the human brain. If this interpretation is correct, the final decisive edge enjoyed by scientific naturalism will come from the capacity to explain traditional religion, its chief competitor, as a

---

66. Dawkins, *Blind Watchmaker*, 287 (italics original).
67. Dawkins, *River out of Eden*, 132.
68. Futuyma, *Science on Trial*, 12.
69. Provine, "Evolution and the Foundation of Ethics," 27.

wholly material phenomenon. Theology is not likely to survive as an independent intellectual discipline.[70]

For Wilson, scientific naturalism is in competition with a religious worldview and seeks to defeat its mortal enemy. Evolutionary biologists are no less motivated by ideological concerns than are ID advocates. Ideological commitment by itself does not necessarily constitute an argument to deny ID the status of science.

Of course, opponents of ID can still argue that ID researchers lack a full understanding of evolutionary biology, rendering ID's criticisms of Darwinian evolution invalid. But this too is a difficult argument to sustain. Consider the educational backgrounds of the some of the leading figures in the ID movement:

- Douglas Axe, PhD in molecular biology, California Institute of Technology
- Michael Behe, Professor of Biochemistry, Lehigh University; PhD in biology, University of Pennsylvania
- David Berlinski, PhD in philosophy, Princeton University; post-doctoral study in mathematics and molecular biology, Columbia University
- William Dembski, PhD in mathematics, University of Chicago; PhD in philosophy, University of Illinois at Chicago
- Stephen Meyer, PhD in the history of biology, Cambridge University
- Scott Minnich, Associate Professor of Microbiology, University of Idaho; PhD in microbiology, Iowa State University
- Lee Spetner, PhD in physics, Massachusetts Institute of Technology
- Jonathan Wells, PhD in biology, University of California at Berkeley; PhD in religious studies, Yale University Divinity School

As a group, ID researchers are well qualified to make scientific judgments on the efficacy of natural selection to account for the diversity of life. One need not agree with their interpretations of the evidence, but one cannot claim they are scientifically uninformed.

According to the grand narrative of Darwinian triumph, ID's status as science was definitively rejected in the 2005 decision by a federal judge

---

70. Wilson, *On Human Nature*, 192.

in the well-publicized *Kitzmiller v. Dover Area School District* case. Many ID advocates testified in this modern-day Scopes Trial, but the judge, John Jones, found ID to be religion dressed up as science and ruled that it had no place being taught in a public school science classroom. But it is not clear how definitive Judge Jones's opinion was.

We saw in chapter 1 that the avowedly atheistic philosopher Thomas Nagel took a surprisingly sympathetic view toward ID in his book *Mind and Cosmos*. But earlier, he had already staked out this position in a criticism of the *Kitzmiller* decision. According to Nagel:

> The political urge to defend science education against threats of religious orthodoxy, understandable though it is, has resulted in a counterorthodoxy, supported by bad arguments, and a tendency to overstate the legitimate scientific claims of evolutionary theory. Skeptics about the theory are seen as so dangerous and so disreputably motivated, that they must be denied any shred of legitimate interest. Most importantly, the campaign of the scientific establishment to rule out ID as beyond discussion because it is not science results in the avoidance of significant questions about the relation between evolutionary theory and religious belief, questions that must be faced in order to understand the theory and evaluate the scientific evidence for it.[71]

Nagel argues that the biological establishment has ruled out the possibility of ID a priori, setting up a double standard. The religious commitment of ID advocates renders ID unscientific, but the naturalistic commitment of the biological establishment does not undermine the scientific status of standard evolutionary theory. Nagel asks the key question:

> Public schools in the United States may not teach atheism or deism any more than they may teach Christianity, so how can it be all right to teach scientific theories whose empirical confirmation depends on the assumption of one range of these views while it is impermissible to discuss the implications of alternative views on the same question?[72]

After all, Nagel observes that while ID will be a more plausible explanation of the evolutionary process to those with a prior religious commitment, it is equally true that Darwinian evolution will seem more plausible to those with a prior materialist commitment. The *Kitzmiller*

---

71. Nagel, "Public Education and Intelligent Design," 187.
72. Ibid., 200.

decision ignored this blatant double standard. ID's status as science will continue to be hotly contested, but what must be realized is that while ID proponents are in fact motivated in part by resistance to what they see as materialist philosophy dressed in the garb of science, this does not mean that they have not produced substantive scientific arguments in support of their theory.

ID theory revolves around three main poles, two that constitute criticisms of Darwinian evolution (irreducible complexity and combinatorial inflation), and one that constitutes a positive argument for design (specified complexity). I will briefly consider each.

Irreducible Complexity

Irreducible complexity is a modern version of William Paley's watchmaker analogy. Based on contemporary work being done in biochemistry and popularized by Michael Behe in the 1990s in *Darwin's Black Box*, the argument from irreducible complexity claims that organisms are made up of complex biochemical machines and processes, at least some of which would not function if any of the individual structures composing them were missing. Like a mousetrap missing a spring or a hammer, these biochemical structures would cease to function if any of their individual parts were removed. The classical structure used to demonstrate irreducible complexity is the bacterial flagellum, the small spinning tail that propels a bacterial cell through a liquid medium. The flagellum is made up of tiny protein structures (a rotor, stator, and bushings) that resemble machines built by humans. If any one of these protein structures is missing, the flagellum ceases to function, leaving the bacterial cell unable to move to find food. Behe argues that a structure like this is irreducibly complex—it cannot be broken down into smaller units—and therefore could not have evolved in a gradual, Darwinian way. Other structures and processes argued to be irreducibly complex include the blood-clotting cascade (a series of biochemical reactions that all need to be performed in the correct order for blood to clot) and the vertebrate eye.

We dealt with the vertebrate eye in chapter 2, but here a few thoughts about the blood clotting cascade are in order. No biologist has spent more time studying this than Russell F. Doolittle. In a recent monograph about the evolution of the vertebrate blood clotting mechanism,

Doolittle argues that the vertebrate blood clotting system must have been assembled in a relatively short period of time since the genes for the clotting mechanism have never been found in the nearest nonvertebrate relatives.[73] While Doolittle traces the appearance of the various clotting factors as they occur in the genomes of various organisms, he does not provide any evidence to prove that a Darwinian process was responsible for the evolution of vertebrate blood clotting, and he freely admits that much of what he does say is speculative.[74] Legitimate questions can therefore be raised about the sufficiency of a gradual Darwinian process to produce this complex series of biochemical reactions that must occur in a precise sequence to prevent us from bleeding to death from a paper cut. The concept of irreducible complexity cannot be automatically ruled out.

In response to the idea of irreducible complexity, Darwinians have countered with the idea of preadaptation. Just because all the individual parts of the bacterial flagellum, for example, are integrated together in a complex formation does not mean that the individual parts originally evolved to be part of the flagellum. The individual parts could have evolved in a Darwinian fashion to perform other tasks, and then later natural selection co-opted them for a new function as parts of the flagellum. That is, the individual parts that came to make up the flagellum were preadapted for their later use as parts of the flagellum. We therefore do not need to posit that the flagellum came into existence all at once with all its parts uniquely designed for this purpose. This argument, however, does seem to fall short. Even if the individual parts of an irreducibly complex system evolved independently for other uses, natural selection would still have had to co-opt them all simultaneously and integrate them into a functional machine. ID proponents do not see how natural selection could do this. For them, irreducible complexity stands as strong evidence against the sufficiency of natural selection to produce complex biological structures.

Combinatorial Inflation

We already met this concept in chapter 4. First raised by MIT mathematicians and engineers at the Wistar Institute conference in 1966, combinatorial inflation refers to the enormous number of ways that the twenty

---

73. Doolittle, *Evolution of Vertebrate Blood Clotting*, 184.
74. Ibid., 197.

amino acids can be arranged to produce protein molecules. Since the evolution of new biological structures would require the evolution of new protein molecules, it is difficult to understand how a process of random mutation could produce functional new proteins, given the exponential number of possible ways amino acids could be arranged in an average protein molecule of three hundred amino acids. The combinatorial possibilities are mind-bogglingly large, on the order of 10^{350} possible combinations! How many of these combinations produce functional proteins rather than gibberish? We know that there are vastly more combinations of the twenty-six letters of the English alphabet that produce gibberish than produce meaningful words. Is it the same for proteins?

Molecular biologist Douglas Axe has focused his research on just this question. He has tried to quantify how many combinations of amino acids would actually produce a functional protein. Can many different combinations lead to functionality, or are functional combinations rare? Surprisingly, Axe has calculated that for our hypothetical protein with three hundred amino acids, there may be as many as 10^{77} functional amino acid arrangements. Yet, despite the enormous size of this number, it still pales in comparison to the number of all possible combinations (10^{350}). Even with such a large number of functional combinations, a random search is highly unlikely to stumble upon even one, given the unimaginably large number of combinatorial possibilities. Axe recounts how he once presented this idea to two prominent experts in the field who felt that Axe's calculations must be in error. Axe relates:

> When the three of us met for a discussion, they had their own hunches about where my reasoning might have gone wrong. Interestingly, though, after perhaps two hours of heated discussion neither agreed with the other's hunch, and we ended up at a polite but dissatisfying impasse. I left with the distinct impression that my conclusion was being rejected not because it was unfounded but because it was unwelcome.[75]

Along with irreducible complexity, ID advocates see combinatorial inflation as further strong evidence against the ability of a Darwinian process to build complex new biological structures. And as Axe points out, this constitutes a *scientific* (not religious) case against Darwinism.[76]

---

75. Axe, "Case against a Darwinian Origin," 11.
76. Ibid.

We should note that this particular criticism of Darwinism has been recognized by the biological establishment—it is not unique to ID proponents or to the mathematicians and engineers who attended the Wistar conference. In 1969, Utah State University biologist Frank Salisbury made the following observation:

> Modern biology is faced with two ideas which seem to me to be quite incompatible with each other. One is the concept of evolution by natural selection of adaptive genes that are originally produced by random mutations. The other is the concept of the gene as part of a molecule of DNA, each gene being unique in the order of arrangement of its nucleotides. If life really depends on each gene being as unique as it appears to be, then it is too unique to come into being by chance mutations. There will be nothing for natural selection to work on.[77]

Salisbury then calculates the number of possible nucleotide sequences in an amino acid protein molecule with three hundred amino acids to be about 10^{600}![78] The problem for Salisbury is apparent:

> In spite of the wild assumptions, the problem should be apparent. In the evolution of life on Earth, we are dealing with millions of different life forms, each based on many genes. Yet the mutational mechanism as presently imagined could fall short by hundreds of orders of magnitude of producing, in a mere four billion years, even a single required gene.[79]

He notes that the problem could be solved by proposing some kind of special creation or directed evolution, but these suggestions, in his opinion, have little scientific value. Yet the idea of purely random mutation cannot overcome the combinatorial problem in Salisbury's view, and he instead speculates on the possibility of some kind of inherent interaction between atoms and molecules that might raise the percentage of useful mutations by many orders of magnitude. Or the existence of feedback mechanisms that might control the order of arrangement of amino acids

---

77. Salisbury, "Natural Selection and the Complexity of the Gene," 342.

78. There are four nucleotide bases. So a protein molecule with three hundred amino acids would contain about one thousand bases (three for each amino acid). Simple combinatorial mathematics shows that a string of one thousand items made up of four items can be arranged in 4^{1000} ways, which equates to about 10^{600}.

79. Salisbury, "Natural Selection and the Complexity of the Gene," 342.

and nucleotides.[80] Salisbury is not the only member of the biological establishment to recognize the problem of combinatorial inflation.

In a 2017 study published in the *Biophysical Journal*, Tian Pengfei and Robert Best attempt to quantify what they term the *sequence capacity* (SC) of a protein molecule, defined as the ability of a protein molecule to withstand changes in its amino acid sequence without experiencing loss of function. Using sophisticated molecular biology techniques, Pengfei and Best measured the sequence capacity of several common protein molecules of varying lengths. They found, as Douglas Axe had, that a protein molecule can perform a specific function with an enormous number of amino acid sequences. Yet these enormous numbers still pale in comparison to all the combinatorial possibilities. For example, a short protein of just thirty-five amino acids has an SC of 10^{22} functional sequences. But the total number of possible sequences is 10^{46}, or one functional sequence for every 10^{24} nonfunctional sequences. For a longer protein of 171 amino acids, the SC inflates to 10^{96}, but the total possibilities inflate to 10^{222}, or one functional sequence for every 10^{126} nonfunctional ones.[81] It is hard to imagine how a random search could find even one new, functional protein even given the high value of the sequence capacity of most protein molecules. The problem of combinatorial inflation clearly is a well-recognized *scientific* criticism of Darwinian evolution.

Specified Complexity

Having provided arguments against Darwinian evolution, ID theorists turn to the concept of specified complexity to make the positive argument for ID as an alternative to Darwinism. Specified complexity is the recognition that some complex structures exhibit features that can be attributed only to a designing intelligence. For example, while hiking in the mountains you might come across some very complex geological formations, but because there are no recognizable patterns, you can safely assume that these formations result from normal physical process like wind and water erosion. But if during your hike you come across a mountain bearing the carved faces of four United States presidents, you would not

---

80. Ibid., 343.
81. Pengfei and Best, "How Many Protein Sequences Fold to a Given Structure?"

ascribe this to wind and water. These patterns must be the work of the intelligent designer of Mount Rushmore.

The concept of specified complexity is regularly employed in a number of fields. Cryptographers use it, as do those searching for extraterrestrial intelligence. Recently archaeologists working in the Amazonian rainforest discovered that buried under the jungle vegetation were a series of carved ditches called geoglyphs. Are these natural formations, or were they constructed by humans? The archaeologists concluded that they were constructed, despite their admission that "surprisingly, little is known about who the geoglyph builders were and how and where they lived, as contemporary settlement sites have not yet been found in the region."[82] Without direct evidence of human activity in the region, the archaeologists still posed a human origin for the geoglyph formations, based on the finding of specified complexity. The geoglyphs look designed, and only human inhabitants would have the intelligence to design them. Similar arguments can be made for the strange rock formations found at Stonehenge or Easter Island.

ID theorists believe biological organisms display evidence of specified complexity. Not least is the fact that biological information is stored in a digital code—DNA—that is not unlike the digital codes—alphabets and computer code—designed by intelligent human beings. As we have already seen, there is no good Darwinian explanation for the origin of DNA or the genetic code. But everywhere we find digital code in our everyday lives, we can trace its origin back to an intelligent designer 100 percent of the time. No computer code has ever come into being via an undirected, unintelligent process. So why shouldn't we view DNA as having originated by an intelligent process too?

Stephen Meyer makes an interesting argument about specified complexity in *Darwin's Doubt*. Meyer, whose educational background is in the history and philosophy of biology, draws on Charles Lyell's uniformitarian approach to geology. Recall that Lyell had transformed geology in the nineteenth century away from the idea of catastrophism toward the notion that the present is the key to the past. That is, to understand how the geological features of the earth were created in the past, we should look to the processes observable today—wind, water, seismic activity, and so forth. Darwin employed the same reasoning in his biological work. To understand how life diversified in the past, we should look to

---

82. Watling et al., "Impact of Pre-Columbian Geoglyph Builders," 1868.

the processes going on in populations today—variation, the struggle for existence, natural selection, and so forth. This is likely why Darwin placed so much emphasis on artificial selection. It was a current, observable process that produced some level of organic change, and in this uniformitarian worldview the present is always the key to the past.

But Meyer turns this around. Given our modern understanding of the complexity of biological organisms and the digitally coded information system standing at their core, Meyer argues that since digital code in our present experience is always attributable to ID, and since the present is the key to the past, it follows naturally that the digital code running biological organisms must also have an intelligent origin. It is the only known explanation of specified complexity. The problem, of course, is that this use of specified complexity requires the existence of a transhuman intelligence, and this is simply not allowed in modern science. But the nonexistence of a supernatural intelligence cannot be established on scientific grounds. It is an idea that can only be held for ideological reasons. So ID may not be a scientific explanation as science is understood today, but denying its scientific status says nothing about its potential truth-value. At the very least, one does not need to accept the idea of ID to realize that ID theorists are basing their views on scientific ideas. As Stephen Meyer says, "Thus, by invoking ID to overcome a vast combinatorial search problem and to explain the origin of new specified information, contemporary advocates of ID are not positing an arbitrary explanatory element unmotivated by a consideration of the evidence."[83]

Despite Meyer's plea, the biological establishment continues to hold ID in contempt and does everything in its power to banish it from the academic world. Any young biologist who openly expresses even the slightest respect for the ID position is guaranteed to be denied research funding and tenure. And those who slip through the cracks in this disciplinary regime, like Michael Behe, are marginalized by their home institutions. Just check the website of the Lehigh University biology department and notice the disclaimer posted there assuring its readers that Behe's ID views are entirely his own and not representative in any way of the views of the biology department or the university. The biological establishment's refusal to engage the ID community along with its attempt to banish ID from the academic world may be backfiring, however. In search of an institutional patron, the ID movement has been driven into

---

83. Meyer, *Darwin's Doubt*, 363.

the arms of the Seattle-based Discovery Institute, a well-funded conservative think tank that is spreading the ID message much more effectively than academia probably could. Through hosting ID design conferences, the dissemination of professionally produced videos, and the publication of ID books from their own press, the Discovery Institute is rapidly spreading the message around the world.

Consider two recent measures of ID's growing influence. In a 2017 edition of the *Journal of Molecular Evolution*, Jan Spitzer published an article reviewing recent work on origin of life research, which he calls "a physicochemical jigsaw puzzle." In his discussion he writes, "Since the subject of cellular emergence of life is unusually complicated (we avoid the term 'complex' because of its association with 'biocomplexity' or 'irreducible complexity'), it is unlikely that any overall theory of life's nature, emergence, and evolution can be fully formulated, quantified, and experimentally investigated."[84] *Biocomplexity* happens to be the title of an ID journal, and we have just seen that irreducible complexity is a central pillar of ID theory. If members of the biological establishment feel it necessary to police their own language in such a way as to not inadvertently employ a common word like *complex* because of its possible associations with ID, then it would appear that ID is gaining the upper hand! This point is strengthened by the announcement in May 2017 that the Discovery Institute has opened a branch at Mackenzie Presbyterian University in São Paulo, Brazil. Mackenzie Presbyterian University is a major Brazilian research university and this is likely the first intelligent-design-oriented entity to be affiliated with a major university anywhere in the world. The strength of the ID movement as an alternative to Darwinian evolution should not be underestimated.

The association of the ID movement with the Discovery Institute does not come without potential costs, however. The Discovery Institute's mission is to advocate for conservative political, economic, and religious viewpoints. Its website speaks of the moral necessity of free markets while standing against the contemporary materialistic worldview, which it says denies dignity and freedom to human beings. This advocacy of free market capitalism tends to align with climate change skepticism. And because ID theory challenges the truth of what many claim to be the most well-established scientific theory of all time—Darwinian evolution—ID proponents associated with the Discovery Institute sometimes

---

84. Spitzer, "Emergence of Life on Earth," 2.

extend their criticism of evolutionary science to science writ large, so they end up as opponents of climate change science too. This is unfortunate as it serves to reify the culture war aspect of the current intelligent-design/Darwinian conflict and takes our eye off the actual scientific issue of accounting for the origin of species. Just because one scientific theory has been perpetuated as a grand narrative does not automatically mean that all have. Each theory needs to be evaluated on its own merits, and anthropogenic climate change is scientifically much better documented than Darwinian evolution. Despite these drawbacks, the ID movement is not going away anytime soon. The scientific establishment would be wise to seriously engage rather than disparage it. As they say, keep your friends close, but your enemies closer.

According to University of Chicago biologist James Shapiro, evolution is best characterized as a continuing mystery whose "fundamental driving forces have not been resolved either in detail or in principle."[85] The discussion in chapters 2 through 7 should have served to demonstrate the general accuracy of this characterization over against the hyperbolic claims of the grand narrative of Darwinian triumph. More than 150 years after the publication of On the Origin of Species, we still do not really know how to account for the stunning diversity of life on Earth. How did life emerge from inanimate matter? How did the first complex cells form? What is the origin of DNA and the genetic code? How do new protein structures evolve given the vast number of combinatorial possibilities? How did animal body plans all form in a short period of time in the Cambrian period, but not before or since? Why do most fossil species change so little over their life histories? Can cells really alter their genomes in intentional ways? How and why did multicellular organisms evolve if bacteria, given their continuing predominance today, were obviously well, and probably better, adapted to the conditions of life? Questions like these and more like them continue to puzzle all who really stop and think about the miracle that is life on Earth. Of course, many theories have been proposed to address these questions, but not a single one of these questions has been given a definitive empirical answer. Despite the best efforts of generations of biologists, the history of life on Earth continues to be a fascinating mystery, and mystery tends to fit better within a religious than a scientific framework.

85. Shapiro, Review of Darwin's Black Box.

What, then, are the implications of this for how we conceive of the contemporary religion–science debate over the issue of the origin and diversity of life? Freed from the need to capitulate to the grand narrative of Darwinian triumph—a narrative designed primarily to advance the interests of biology as a proper scientific discipline—how might we reconceive of the religion–science relationship in a way that does full justice both to the amazing insights about life that biology *has* been able to produce and to the necessarily more subjective nature of religious reflection? What positive role, if any, can religious reflection play in discerning objective truth about the world? It is to these big questions that I turn in the final chapter.

8

Rethinking the Religion–Science Relationship

IF THERE WAS EVER a question about the true motivation for maintaining the materialist foundations of modern biology, it was put to rest by Harvard geneticist Richard Lewontin in 1997 in this oft-quoted passage from a review of Carl Sagan's *The Demon-Haunted World*:

> Our willingness to accept scientific claims that are against common sense is the key to an understanding of the real struggle between science and the supernatural. We take the side of science in spite of the patent absurdity of some of its constructs, in spite of its failure to fulfill many of its extravagant promises of health and life, in spite of the tolerance of the scientific community for unsubstantiated just-so stories, because we have a prior commitment to materialism. It is not that the methods and institutions of science somehow compel us to accept a material explanation of the phenomenal world, but, on the contrary, that we are forced by our a priori adherence to material causes to create an apparatus of investigation and a set of concepts that produce material explanations, no matter how counterintuitive, no matter how mystifying to the uninitiated. Moreover, that materialism is absolute, for we cannot allow a Divine Foot in the door.[1]

As we have seen, some of the leading voices of the biological establishment over the last thirty years have joined Lewontin in holding this a priori commitment to materialism, leading to a dogmatic adherence to

---

1. Lewontin, "Billions and Billions of Demons."

Darwinian evolution in order to keep the Divine Foot outside the door. But as the discussion of the last seven chapters has shown, Darwinian evolution, while it may be properly scientific in its commitment to materialist explanation, is about as empirically uncertain as any widely accepted scientific theory could possibly be. It is therefore not necessary for religion scholars who study the relationship between religion and science to begin their reflections on that relationship with an uncritical nod to Darwinism. The field of evolutionary theory is actually wide open. What are the implications of this for religion scholars?

Over the last couple centuries, religion and science have come, in the popular imagination, to be constructed on a rigid objective/subjective binary. Science is thought to be rigorously empirical and objective. Scientific theories are based on the observation, measurement, and interpretation of data from the physical world of matter and energy, not on the preconceived biases of the scientists. Scientific results must be replicable to be taken seriously, and rigorous peer review of these results ensures that these conventions of objectivity are maintained. Religion, on the other hand, is considered to be subjective and therefore nonempirical. Religious ideas are accepted on faith and their meaning open to a multiplicity of interpretations. Religion is personal and experiential, and deals by definition with nonmaterial entities like gods, spirits, ethics, and values.

In addition, the spread of materialist philosophy in the modern world has led to a prioritization of the physical over the spiritual so that scientific investigation is now seen by many as the only acceptable avenue toward truth. If the world is made of nothing more than physical stuff (elementary particles, atoms, molecules, and their interactions), then science, the mode of investigation that objectively deals with this physical stuff, must be the only way to understand the nature of the reality in which we live. That is, science is not only objective, but it deals with real things. Religion, by contrast, is rendered unreal. If nonphysical entities like gods or spirits are considered unreal by virtue of their immateriality, then religion can be marginalized as essentially fictional. Though some might still see religion as a useful fiction in its role to help order human society and produce ethical norms, useful or not, religion is rendered fictional—it does not deal with the physical entities that make up the real world. Science, therefore, becomes by default the sole arbiter of objective truth.

Once religion and science are placed in these rigid objective–subjective and real–unreal binaries, the conclusion follows that religion and science cannot, by definition, be in conflict. Since they deal with entirely different realms of human understanding and experience, there are no grounds for conflict between them. This idea has been most famously expressed by Stephen Jay Gould with his concept that religion and science constitute nonoverlapping magisteria (or NOMA). According to Gould:

> The lack of conflict between science and religion arises from a lack of overlap between their respective domains of professional expertise—science in the empirical constitution of the universe, and religion in the search for proper ethical values and the spiritual meaning of our lives.[2]

Gould believes both science *and* religion are important, and he appeals to scientists and religious folk alike to respect the mutually exclusive nature of religion and science:

> I believe, with all my heart, in a respectful, even loving, concordat between our magisteria—the NOMA solution. NOMA represents a principled position on moral and intellectual grounds, not a mere diplomatic stance. NOMA also cuts both ways. If religion can no longer dictate the nature of factual conclusions properly under the magisterium of science, then scientists cannot claim higher insight into moral truth from any superior knowledge of the world's empirical constitution. This mutual humility has important practical consequences in a world of such diverse passions.[3]

If only it were so simple. We have already seen how the very scientists who criticize ID advocates for crossing these boundaries and injecting religion into their scientific work themselves cross these boundaries all the time, drawing sweeping philosophical, theological, and ethical conclusions from evolutionary theory—for instance when, as direct conclusions from Darwinian evolution William Provine declares ethics a worthless pursuit, and Douglas Futuyma calls human life meaningless. The NOMA solution is attractive on the surface but difficult to maintain in practice.

2. Gould, "Nonoverlapping Magisteria," 2.
3. Ibid., 8.

To illustrate this, consider the editorial comments appearing in Francisco Ayala and John Avise's *Essential Readings in Evolutionary Biology*. In one place they seem to channel Gould:

> The scope of science is the world of nature, the reality that is observed, directly or indirectly, by our senses. Science advances explanations concerning the natural world: the composition of matter, the drift of the continents, the expansion of the galaxies, the origin and adaptation of organisms. Religion concerns the meaning and purpose of the world and of human life, the proper relation of people to the Creator and to each other, the moral values that inspire and govern people's lives. It is only when assertions are made beyond their legitimate boundaries that science and religious belief appear to be antithetical.[4]

But then in another place, Ayala and Avise comment on biochemical flaws in the human body and ask, "why do biological flaws exist in a world supposedly run by an omnipotent and beneficent supernatural designer?"[5] They then state, "This scientific revelation is but one of many reasons why the findings of evolutionary biology are highly relevant to philosophical discourse on numerous topics traditionally reserved for theologians and other religious practitioners."[6] If Ayala and Avise believe that scientific findings have a direct bearing on theological issues, then they would appear to be transgressing the very boundary between science and religion that they say they want to defend! NOMA is not a convincing solution to the religion–science "conflict." It is espoused mainly by scientists who desire to maintain the materialist purity of science, maintain their hegemony over objective truth, and keep religion out of the laboratory. But reality does not come to us in such hermetically sealed compartments.

A more realistic assessment of the relationship comes from Harvard astronomer Owen Gingerich in his little book *God's Planet*:

> The bottom line that will emerge from these three chapters is that science, working within its own magisterium, is far more tangled with a humanistic or theological vision than we might expect, and that the magisteria are more overlapping than we might idealistically (from a strictly scientific perspective) suppose.[7]

4. Ayala and Avise, eds., *Essential Readings in Evolutionary Biology*, 273.
5. Ibid., 44.
6. Ibid.
7. Gingerich, *God's Planet*, 29.

Based on this view, Gingerich offers the following sage advice: "If someone tells you that evolution is atheistic, be on guard. If someone claims that science tells us we are here by pure chance, take care. And if someone declares that magisteria do not overlap, just smile smugly and don't believe it."[8] If the magisteria do in fact overlap, religious ideas may have as many implications for scientific ideas as scientific ideas have for religious ones. It may not be the case that religion must always be subordinated to science even in reference to questions about the nature of physical reality and especially in reference to questions about the origin and diversity of life.

If Gingerich is right, religious thought may pose a real challenge to Darwinian evolution as a naturalistic scientific theory, as E. O. Wilson has readily admitted: "Religion constitutes the greatest challenge to human sociobiology and its most exciting opportunity to progress as a truly original theoretical discipline."[9] If religion poses this kind of challenge, it will need to be vanquished. Thus, Wilson argues that "the principle of natural selection acting on the genetically evolving material structure of the human brain" can explain religion. Thus "the final decisive edge enjoyed by scientific naturalism will come from its capacity to explain traditional religion, its chief competitor, as a wholly material phenomenon. Theology is not likely to survive as an independent intellectual discipline."[10] Wilson wrote these words in 1978, but theology (even in the hallowed halls of Wilson's Harvard) has continued to flourish as an independent intellectual discipline for the forty years since. Perhaps Wilson considerably overstated the power of Darwinism to explain religion away. Despite its threat to the naturalistic foundations of science, religious thought and experience cannot be ignored as a potential avenue toward objective truth about the world. There is no need for religion scholars and theologians to cede that ground entirely to the scientists. The objective–subjective binary must be critically examined. This is what I will do in this final chapter.

---

8. Ibid., 153.
9. Wilson, *On Human Nature*, 175.
10. Ibid., 192.

Problematizing the Objective–Subjective Binary

Despite the close association between Darwinian evolution and naturalistic philosophy that leads high-profile Darwinians to draw broad atheistic conclusions from Darwinism, not all evolutionary biologists are atheists. Many biologists find no conflict between their support for Darwinian evolution and a traditional religious worldview. Ironically, creationists and atheistic scientists—enemies to the core—do find unexpected common ground on the issue of the incompatibility of religion and Darwinian evolution; both see them as incompatible. But many others do not and find no problem with professing both a traditional faith in the existence of God and a belief in the sufficiency of natural selection to account for the evolution and diversity of life. This view is usually termed *theistic evolution*. Theistic evolution is a complex phenomenon whose full treatment is beyond the scope of this book. But I will briefly consider here the work of three biologists who fall into this category (Kenneth Miller, Francis Collins, and once again Theodosius Dobzhansky) to determine whether theistic evolution is an intellectually coherent option that fully supports the objective–subjective binary. If theistic evolution proves to be intellectually incoherent, the binary may have to be rethought.

Kenneth Miller

In 1999, Brown University cell biologist Kenneth Miller published *Finding Darwin's God*, a book that presents Miller's view on how to square a firm commitment to Darwinian evolution with his conservative Roman Catholic faith. Miller seems to be a proponent of NOMA, but in articulating his understanding of the different roles played by science and religion, he makes the following interesting comment:

> By definition, a god is a nonmaterial being who transcends nature, so why should science, which deals only with the material world, have anything to say about whether or not a god exists? In the rigorous, logical sense, it shouldn't. But we are a practical species, interested in getting results. Humans like to feel that their beliefs have a link to reality, and here's where science has it all over religion.[11]

---

11. Miller, *Finding Darwin's God*, 194.

Of course, humans *do* like to feel that their beliefs have a link to reality. But if this is where science has it all over religion, Miller seems to imply that religion does not deal with reality, only science does. He then explains further:

> We can light a candle to explain the commonplace activities of everyday life, by showing that an understandable, *material* mechanism is at work in each of them—in short, by showing that the phenomenon at hand is a property of the ordinary stuff of nature. *That* is the working assumption of materialism—namely that nature itself is where we can find the explanations for how things work. It is also the credo of science—making science, by definition, a form of practical, applied materialism.[12]

And then in another place Miller repeats for emphasis, "Where theology promises only miraculous explanations for nature, science can show you the *real* explanation."[13]

What exactly does Miller's God do, or how is this God made manifest in the world if all physical phenomena can by definition be given a physical (a real) explanation? Miller does not seem bothered by this question. For him, evolution does nothing to weaken the power of God. In fact:

> A God who presides over an evolutionary process is not an impotent, passive observer. Rather, He is one whose genius fashioned a fruitful world in which the process of continuing creation is woven into the fabric of matter itself. He retains the freedom to act, to reveal Himself to His creatures, to inspire, and to teach. He is the master of chance and time, whose actions, both powerful and subtle, respect the independence of His creation and give human beings the genuine freedom to accept or reject His love.[14]

But how does Miller's God do all this? How is God revealed to God's creatures when the origin and evolution of God's creatures can be given a completely physical explanation that Miller asserts is more real than a religious explanation? If one cannot see God acting in nature (the real world), then how can one know with any assurance that God presides over the evolutionary process in the active way that Miller imagines?

---

12. Ibid. (Italics original).
13. Ibid., 209 (italics original).
14. Ibid., 243.

To solve this conundrum, Miller improbably appeals to the principle of quantum indeterminacy. Because events that occur at the level of elementary particles are unpredictable and indeterminate, Miller argues that this would allow a "clever and subtle God to influence events in ways that are profound, but scientifically undetectable to us."[15] He includes events like the appearance of mutations, the activation of neurons in the brain, or the survival of cells or organisms affected by radioactive decay. So the religious Miller wants to see God directly at work in the physical world through the agency of quantum indeterminacy, while the scientific Miller wants to make sure that God's activity remains scientifically undetectable. But by emphasizing that science provides the real explanation for things, Miller seems to render religious faith something unreal and purely subjective. How many religious believers will be happy with a theistic evolution that renders God unreal and therefore faith in the existence of God a purely subjective hope? If God, gods, or some kind of creative intelligence is truly active in the world, why should it not be scientifically detectable? Wouldn't this be a reasonable expectation? The intellectual coherence of Miller's brand of theistic evolution appears dubious.

More to the point, why is Miller comfortable positing the direct action of God in a scientific discipline not his own (quantum physics) but not in his own discipline of biology? Who is Miller to restrict God's activity to the realm of elementary particles while banishing it from the realm of molecules, cells, or organisms?

Francis Collins

Francis Collins is the director of the National Institutes of Health and earlier was a leader of the Human Genome Project, a massive effort completed in 2003 to map the human genome. He is also an avowed evangelical Christian and in 2005 published *The Language of God: A Scientist Presents Evidence for Belief*. Like Miller, Collins tries to square his acceptance of Darwinian evolution with his traditional Christian beliefs. And also like Miller, he makes some curious comments in the process.

Collins begins with a NOMA-like articulation of the relationship between science and religious faith:

---

15. Ibid., 241.

> If God exists, then He must be outside the natural world, and therefore the tools of science are not the right ones to learn about Him. Instead, as I was beginning to understand from looking into my own heart, the evidence of God's existence would have to come from other directions, and the ultimate decision would be based on faith, not proof.[16]

For Collins, evidence for God should not be sought within events occurring in the physical world. This is all well and good except that Collins, like Miller, later contradicts himself when commenting on modern cosmology:

> The Big Bang cries out for a divine explanation. It forces the conclusion that nature had a defined beginning. I cannot see how nature could have created itself. Only a supernatural force that is outside of space and time could have done that.[17]

Many modern cosmologists would vehemently disagree with Collins here, arguing that in fact the universe might well be self-creating—the result of a random quantum vacuum fluctuation. Be that as it may, the important point here is that Collins also wants to keep religious explanations out of his domain of biology but seems only too willing to pose religious explanations for the domains of physics and cosmology. He cannot have it both ways. If God acts directly in the realm of physics, on what grounds can anyone say that God does not act directly in the biological realm too?

Collins goes on to describe surprises that he encountered through the process of unravelling the human genome. One of the biggest surprises was finding that the human genome contains only about twenty thousand to twenty-five thousand protein-coding genes, accounting for a measly 1.5 percent of all the DNA in the genome. Collins quips, "After a decade of expecting to find at least 100,000 genes, many of us were stunned to discover that God writes such short stories about humankind."[18] One would expect that as a committed Darwinian Collins would understand the human genome as having arisen as a result of a long process of random mutation and natural selection, not as something that is in some way designed by God to be as short as it is. Is God directly involved in a physical process like the evolution of the human genome or not?

16. Collins, *Language of God*, 30.
17. Ibid., 67.
18. Ibid., 124.

Countering Richard Dawkins's notion that Darwinian evolution fully accounts for biological complexity, leaving no room for God, Collins writes, "While this argument rightly relieves God of the responsibility for multiple acts of special creation for each species on the planet, it certainly does not disprove the idea that God worked out His creative plan by means of evolution."[19] Collins here seems to conveniently ignore the basic premise of Darwinian evolution: the evolutionary process is unplanned and undirected. Dawkins's understanding may be wrong in Collins's view, but at least it is intellectually coherent. Collins, by positing divine activity as driving an unplanned and undirected process, seems considerably more muddled.

But undeterred, Collins concludes:

> I found this evidence of the relatedness of all living things an occasion of awe, and came to see this as the master plan of the same Almighty who caused the universe to come into being and set its physical parameters just precisely right to allow the creation of stars, planets, heavy elements, and life itself. Without knowing its name at the time, I settled comfortably into a synthesis generally referred to as "theistic evolution," a position I find enormously satisfying to this day.[20]

Collins may find theistic evolution satisfying, but how many others will? If the events of the physical world can be completely explained naturalistically by scientists like Collins, what does this Almighty actually do? And how can we know that this Almighty is real if there are no physical signs of its existence? Collins's position seems no more intellectually coherent than Miller's.

Theodosius Dobzhansky

As we saw earlier, this great hero of the grand narrative of Darwinian triumph was the most traditionally religious of the leading figures of the modern synthesis. A committed Russian Orthodox Christian who had great respect for the evolutionary theology of Pierre Teilhard de Chardin, Dobzhansky can certainly be characterized as a proponent of theistic evolution. It will be helpful in this regard to consider two additional

---

19. Ibid., 163.
20. Ibid., 199.

interesting points encountered in Dobzhansky's oft-cited "Nothing in Biology Makes Sense Except in the Light of Evolution."

One of the principal arguments atheistic Darwinians make against the notion that God either directly created or intelligently designed the organic world revolves around the many apparent flaws in nature. We have already seen this in terms of the inverted retina of the vertebrate eye. Why would an intelligent designer design things this way? As another example, consider this passage from ardent Darwinian Jerry Coyne:

> Why would a creator happen to leave amphibians, mammals, fish, and reptiles off oceanic islands, but not continental ones? Why did a creator produce radiations of similar species on oceanic islands, but not continental ones? And why were species on oceanic islands created to resemble those from the nearest mainland? There are no good answers—unless, of course, you presume that the goal of a creator was to make species *look* as though they evolved on islands.[21]

It is a common rhetorical ploy to ask why an intelligent creator would do the inexplicable things it did as way to offer Darwinian evolution as a de facto superior alternative. This is fine as long as one is working from an atheistic framework. But what if one isn't?

Dobzhansky certainly wasn't, yet he employed this same kind of argument when commenting on the phenomenon of extinction. It is well known that about 99 percent of all species that have ever lived on Earth have gone extinct. According to Dobzhansky, "All this is understandable in the light of evolution theory; but what a senseless operation it would have been, on God's part, to fabricate a multitude of species ex nihilo and then let most of them die out!"[22] Such an argument against creationism might work for an atheist like Coyne, but how can a theist like Dobzhansky make it? He boldly proclaims, "I am a creationist *and* an evolutionist. Evolution is God's, or Nature's, method of Creation."[23] So maybe God did not create all organisms ex nihilo, but God still, on Dobzhansky's logic, created an evolutionary process that has led to the extinction of almost all the species that the process has ever created. Dobzhansky's use of the God-wouldn't-do-it-this-way argument lacks coherence. For in Dobzhansky's theistic worldview, God manifestly did do it this way!

21. Coyne, *Why Evolution Is True*, 108 (italics original).
22. Dobzhansky, "Nothing in Biology," 126.
23. Ibid., 127.

On this mode of argument, the reflections of the famous quantum theorist Niels Bohr are in order. As reported by Werner Heisenberg, Albert Einstein was famously averse to the statistical nature of quantum theory; his aversion summed up in his well-known dictum "God does not play dice." To this Bohr is said to have responded, "Nor is it our business to prescribe to God how He should run the world."[24] Deciding what an intelligent creator would or would not do is not a particularly compelling argument in favor of an undirected Darwinian process. But at least it is an intellectually coherent argument when made by atheistic biologists. It loses that coherence, however, when made by a theistic evolutionist like Dobzhansky.

Further troubles follow when Dobzhansky rehearses an orthodox Darwinian line: "There is, of course, nothing conscious or intentional in the action of natural selection. A biologic species does not say to itself, 'Let me try tomorrow (or a million years from now) to grow in a different soil, or use a different food, or subsist on a different body part of a different crab.'"[25] There is nothing unusual about this statement. Natural selection has no larger purpose or end goal. But then Dobzhansky follows with, "Only a human being could make such conscious decisions. This is why the species *Homo sapiens* is the apex of evolution."[26] Apex of evolution?! In what sense can an undirected process with no larger purpose ever be described as moving toward an apex? Seeing humans as the crowning glory of creation certainly fits with Dobzhansky's Christian sensibilities, but it is diametrically opposed to his attempt to also hold to orthodox Darwinian sensibilities. If evolution *is* an undirected process, then humans can never be understood as the apex of that process, for by definition an undirected process has no apex. Dobzhansky's attempt to articulate a theistic evolutionary position falls once again into intellectual incoherence.

Theistic evolution would not seem to be a compelling solution to the religion-evolution relationship. By uncritically accepting the objective–subjective binary, theistic evolution essentially reifies the idea that science objectively works with the real world, tacitly rendering religious ideas purely subjective and unreal. One can assert God as a creator, but such an assertion is drained of any real meaning if at the same time one

---

24. Cited in Heisenberg, *Physics and Beyond*, 81.
25. Dobzhansky, "Nothing in Biology," 127.
26. Ibid.

argues that the origin and diversity of life can be fully explained by a purely material process. As I have tried to show throughout this book, the idea that the origin and diversity of life can be fully explained in this way is a rhetorical device designed to support the guild interests of the scientific establishment. The question of life's origin and diversity is actually far more mysterious. Once again, Werner Heisenberg is most helpful here:

> The language of religion is more closely related to the language of poetry than to the language of science. True, we are inclined to think that science deals with information about objective facts, and poetry with subjective feelings. Hence we conclude that if religion does indeed deal with objective truths, it ought to adopt the same criteria of truth as science. But I myself find the division of the world into an objective and a subjective side much too arbitrary. The fact that religions through the ages have spoken in images, parables, and paradoxes means simply that there are no other ways of grasping the reality to which they refer. But that does not mean that it is not a genuine reality.[27]

This reality-unreality binary finds expression also in the debate over metaphysical and methodological materialism (or naturalism). Some theistic scientists argue that as scientists they are duty bound to search for material explanations in their scientific research, a position known as methodological materialism. But this does not mean that they necessarily ascribe to metaphysical materialism, the notion that reality itself is composed of nothing but atoms, molecules, and energy. That is, they can work within a materialist framework in their scientific methodology, but still affirm a belief in God, gods, or some spiritual reality in their overall worldview. One such theistic scientist is Robert Pennock, who puts it this way:

> Methodological naturalism does not define away any metaphysical possibilities or constrain the world, rather, it constrains science. Methodological naturalism is neutral with regard to supernatural possibilities. It takes a more humble view of what can be known. Science admits that it may miss true metaphysical facts about the world. Methodological naturalism does not claim access to all possible truths. Indeed, it expressly limits the purview of what can be known scientifically. If there are

---

27. Heisenberg, *Physics and Beyond*, 87.

metaphysical truths beyond empirical test, then they are beyond science.[28]

This distinction between metaphysical and methodological materialism allows Pennock to remain a scientist in good standing and simultaneously a religious believer. But at what cost? Does not adhering to methodological materialism turn out to be a form of de facto metaphysical materialism? If a scientist enters into investigation of the physical world with the assumption that a material explanation of the phenomenon she is studying must exist, she will miss any nonmaterial explanations that might actually be occurring.

ID advocate William Dembski makes an important point in response to Pennock:

> It follows that the charge of supernaturalism against intelligent design cannot be sustained. Indeed, to say that rejecting naturalism entails accepting supernaturalism holds only if nature is defined as a closed system of material entities ruled by unbroken laws of material interaction. But this definition of nature begs the question. Nature is what nature is, not what we define it to be.[29]

Dembski correctly understands the nature-supernature binary itself to be an artificial and arbitrary linguistic construction designed to maintain a purely materialist view of science. If some immaterial reality exists in the universe, it is just as much a part of reality as the material stuff that scientists study, and it may have direct causative effects on that material stuff. Reality is what reality is. We don't get to define it according to our purposes and interests. To adhere to methodological materialism becomes then a sort of de facto adherence to metaphysical materialism. This is fine as long as metaphysical materialism is an accurate description of reality. But is there scientific evidence to support a metaphysically materialist worldview?

---

28. Pennock, "The Pre-Modern Sins of Intelligent Design," 739.
29. Dembski, "In Defense of Intelligent Design," 727.

How Material is Materialism?

"Consciousness is the most conspicuous obstacle to a comprehensive naturalism that relies only on the resources of physical science."[30] With these words, philosopher Thomas Nagel attempts to deconstruct a materialist, Darwinian understanding of the natural world. Nagel goes on, "Materialism is incomplete even as a theory of the physical world, since the physical world includes conscious organisms among its most striking occupants."[31] For Nagel as for many other thinkers, the phenomenon of conscious awareness, so powerful a part of human experience and seemingly for many other types of organisms, undermines any attempt to hold to a comprehensive materialist view of the nature of reality. The thoughts and feelings that run through our conscious awareness during our every waking minute (and in dreams while we are asleep) are by definition immaterial. One cannot weigh or measure a feeling. One cannot touch, smell, or taste a thought. What are often referred to as the qualia of conscious awareness are among the most mysterious aspects of biological organisms, and at least in their human form, among the most powerful.

Left to our physical biological endowment, we humans would have a limited impact on the physical environment. We are small, weak, and slow, and have poor eyesight and a poor sense of hearing and smell compared to other creatures, a point made well by Derrick Jensen in *The Myth of Human Supremacy*. But with the power of a sophisticated immaterial mind, humans have completely altered the face of the planet. No skyscraper, dam, or superhighway was ever constructed that did not first exist as an immaterial idea in the mind of an architect or engineer. No great city or global economic system ever came into being apart from the direct action of immaterial conscious minds. Every scientific theory—even one like Darwinian evolution that proposes a purely material explanation for the mind—is, ironically, itself the product of the immaterial mind! With minds we can even alter the evolutionary process itself through genetic engineering. The mind is a real and powerful force in the world, but can it be explained materially? Nagel does not think so, and he is far from alone. Others, of course, will disagree.

Many philosophers and cognitive scientists like Daniel Dennett have tried to reduce mind to the physical actions of a material brain, but so far with very little success. In a 2011 article, evolutionary psychologist

30. Nagel, *Mind and Cosmos*, 35.
31. Ibid., 45.

David Barash proclaimed the question of the nature of consciousness to be the hardest problem in science. For Barash,

> It's a hard one indeed, so hard that despite an immense amount of research attention devoted to neurobiology, and despite great advances in our knowledge, I don't believe we are significantly closer to bridging the gap between that which is physical, anatomical, and electro-chemical, and what is subjectively experienced by all of us.[32]

In fact, according to Barash, the hard problem of consciousness is so hard that he can't even imagine what kind of empirical findings might solve it. He is not even sure what kind of discovery could get us to first base. And yet, Barash confidently proclaims:

> I write this as an utter and absolute, dyed-in-the-wool, scientifically oriented, hard-headed, empirically insistent, atheistically committed materialist, altogether certain that matter and energy rule the world, not mystical abracadabra. But I still can't get any purchase on this "hard problem," the very label being a notable understatement.[33]

Barash is absolutely certain that a material explanation exists even though he cannot even conceive of what that explanation might look like. If only other Darwinians were so wonderfully transparent about their philosophical biases!

Minds are real and appear to be among the most powerful shaping forces in the world (and possibly in the universe). But their origin remains mysterious and seemingly beyond the grasp of material explanation. Even the population geneticist Sewall Wright felt the need to resort to the idea of panpsychism to explain the existence of mind. So metaphysical materialism would appear to be a dead end as a comprehensive understanding of the nature of reality. Of course, this is not a new idea. Hindu, Buddhist, and Sufi meditators have been exploring the vast dimensions of human consciousness directly for millennia. It comes as no surprise to them that the mind cannot be reduced to mere material causation. Western science seems to be finally catching up.

Recent years have seen many scientists beginning to take seriously various forms of parapsychological experience like past-life remembrance, near-death experience, out-of-body experience, and extrasensory

---

32. Barash, "Hardest Problem in Science?"
33. Ibid.

perception, seeing in these experiences valid evidence for the mind as existing independent of the physical body. In 2015, Edward Kelly, research professor in the Department of Psychiatry and Neurobehavioral Sciences at the University of Virginia, coedited a volume of essays titled *Beyond Physicalism: Toward Reconciliation of Science and Spirituality* that provides a rigorous scientific and philosophical approach to parapsychological phenomena. Also in 2015, Marjorie Hines Woollacott, a longtime neuroscience professor at the University of Oregon, published *Infinite Awareness: The Awakening of a Scientific Mind*. Woollacott, who not only works as a scientist, but who has long practiced meditation, has reached the conclusion as well that the mind cannot be reduced simply to the functioning of the physical brain. And mention should also be made of Jo Marchant's 2016 *Cure: A Journey into the Science of Mind over Body*, a study of the placebo effect and other ways that the mind is being harnessed directly by medical researchers to produce actual physical changes in the body. Scientists are quickly discovering that the reality of the immaterial mind cannot be ignored in any comprehensive theory about the nature of the universe. The universe appears to be more than physical stuff.

But the materiality of the physical stuff itself—particles, atoms, and molecules—is also being called into question by the science of quantum theory. In the words of Noson Yanofsky:

> Probably the greatest development in all of physics is quantum mechanics. With the exception of gravity, all physical phenomena are described by this theory... However, quantum mechanics has also taught us that we have a severe limitation when it comes to understanding how the particles of our universe behave. They are extremely mysterious and defy our attempts to make sense of them.[34]

As we have known for over a century, elementary particles like electrons cannot be assigned either to the category of matter or energy. If one designs an experiment to demonstrate that electrons act like small particles of matter, they will act like small particles of matter. But if one designs an experiment to demonstrate that electrons act like waves of energy, they will act like waves of energy. This wave-particle duality, which defies common sense, is a fundamental feature of all elementary particles. Not to mention that all elementary particles have no definite location until

---

34. Yanofsky, *Outer Limits of Reason*, 175.

they are measured. They exist in a state of superposition described by a probability wave that can tell us only the odds of finding the particle in any particular location. Only when the particle is observed by a scientist does the probability wave collapse to a definite value. The quantum world is more bizarre than *Alice in Wonderland*. But there is more.

The famous Heisenberg uncertainty principle tells us that it is impossible to measure both the location and momentum of an elementary particle precisely at the same time. The more precise our determination of its location, the less precise will be our knowledge of where it is going. The more precise our measurement of where it is going, the less precise will be our knowledge of where it is! The very act of measuring these quantities changes them. The mind is not a passive observer of a physical process going on out in the world. The mind actively engages with the quantum world to determine its future unfolding. And that unfolding is inherently uncertain and unpredictable. With a ball, if you know its location and momentum (speed and direction), you can accurately predict where it will be at any future instant. But this determinism is impossible at the level of elementary particles. Even the most powerful computer in the world could not by definition predict the future course of the evolving cosmos. Uncertainty and indeterminism are fundamental features of our reality.

Even more bizarre than quantum indeterminacy is quantum entanglement. When a pair of elementary particles are created in some physical reaction, they remain mysteriously entangled with each other in such a way that one particle seems to "know" what is happening to the other particle instantaneously no matter how far apart they might be. For example, if we create a pair of electrons and let them fly off in opposite directions, we know that a property of electrons (e.g., the property that electrons spin) exists in a state of superposition. Each electron can only be described by a probability equation that tells us the likelihood of the electron having any particular value of spin. So as our pair of electrons travel apart from each other, experiments have shown that if we measure the spin of one electron (thus collapsing its probability wave), the other electron's probability wave collapses simultaneously to the precise value needed to maintain the conservation of spin. That is, the second electron seems to "know" instantaneously what happened to the first electron and acts accordingly! Einstein, no fan of quantum theory, derided entanglement as "spooky action at a distance" and tried to find a more "classical" explanation for the phenomenon. But to no avail. In our normal life experience, information can never flow faster than the speed of light and is always mediated by some source of energy (light waves, sound waves, radio waves, and so forth). But not so in the quantum world.

Commenting on the phenomenon of quantum entanglement, Yanofsky writes:

> One consequence of entanglement is to end the philosophical position of reductionism. This position says that if you want to understand some type of closed system, look at all the parts of the system. To understand how a radio works, one must take it apart and look at all of its components, because "the whole is the sum of its parts." Reductionism is a fundamental supposition in all of science. Entanglement shows that there are no closed systems. Every part of a system can be entangled with other parts outside of the system. All different systems are interconnected and the whole universe is one system. One cannot understand a system without looking at the whole universe. That is, "the whole is *more than* just the sum of its parts."[35]

Once again, modern science is pushing us back to insights registered millennia ago by Buddhist philosophers, the essential interconnectedness of all things.

The materials at the heart of materialism—elementary particles—do not appear to be very material at all. The physical world in which we live fuzzes out into a haze of quantum indeterminacy at the level of elementary particles. Add to this the reality of the immaterial mind as an essential feature of reality, and there is really no longer a good argument to be made for holding to a strictly materialist and reductionist philosophy. The universe is a far more mysterious place than we might generally realize.

The assumption of material causation for all physical phenomena stands now at the heart of scientific inquiry, but the wall built between objective science and subjective religion is a modern construction. Some of the greatest of Western scientists prior to the eighteenth century interwove religion and science in ways that would make many scientists today cringe.[36] For example, what exactly led Copernicus to propose a radical new idea for the structure of the cosmos? Was he driven to put the sun at the center of the universe by incontrovertible physical evidence?

---

35. Ibid., 201.

36. For an account of the shifting interpretations of the terms *religion* and *science*, see Harrison, *Territories of Science and Religion*.

Not at all. Owen Gingerich remarks, "As the seventeenth century began there was no physical evidence in favor of the heliocentric system beyond its unifying aesthetic."[37] Evidence did not motivate Copernicus; rather a religiously oriented aesthetic motivated him.

By Copernicus's time Ptolemy's Earth-centered universe had grown complicated, buttressed by ad hoc features meant to account for inexplicable qualities that astronomers kept observing. Ad hoc features attached to the Ptolemaic system included *epicycles* (to account for the retrograde motion of some planets) and *equants* (to account for the speed changes in the orbits of some planets). Such features had to be added to the Ptolemaic system every time astronomers made new observations that did not fit the system's expectations. In Bruce Wrightsman's view:

> Copernicus could not believe that the God whom he affirmed in *De Revolutionibus* as the 'universal Artisan of all things' and the 'Best and most orderly workman' would be so clumsy as to have created such a monstrosity. It was thus on the basis of his belief in a Creator God who was the 'Best and Greatest Artist' that he sought for 'purer and more convenient assumptions' that would be consistent with the principle of uniform motion.[38]

Copernicus realized that putting the sun at the center of the cosmos led to a much simpler and more elegant solution to the question of the structure of the observable universe. The Copernican system could account for observation without recourse to ad hoc features. Copernicus had no empirical evidence to support this radical transformation of worldview; indeed the complicated Ptolemaic system still worked well as far as making accurate predictions about the movement of planets and stars. Copernicus's motivation was primarily theological. In contrast to today's atheists who appeal to a God-wouldn't-do-it-this-way argument in their criticism of ID, Copernicus appealed to a God-would-do-it-this-way argument in support of his radically transforming worldview.

Yet problems with Copernicus's heliocentric model remained. By putting the sun at the center of the cosmos and allowing for the earth to revolve around it, Copernicus theorized that the motion of the sphere of the fixed stars must only be apparent motion, caused by the motion of the earth. The sphere of the fixed stars must be immobile. Yet in Copernicus's view, it is natural for a sphere to rotate. So how is it possible for

---

37. Gingerich, *God's Planet*, 41.
38. Wrightsman, "Legitimation of Scientific Belief," 55.

the sphere of the fixed stars to remain immobile? Copernicus appeals to the Platonic-Pythagorean tradition holding that immobile heavenly objects are nobler than those that move. He wrote, "We conceive immobility to be nobler and more divine than change and instability, which latter are more appropriate to earth than to the universe."[39] Once again, Copernicus resolves the dilemma by appealing to theological justification in a way that would be considered anathema in contemporary science. Wrightsman concludes:

> One may not like Copernicus's reasons for coming to believe in and justifying his system but that is not a rational ground for refusing to accept them as *reasons*. We must therefore remind ourselves that scientific investigation had much broader implications for Copernicus than it has for many today, and included those purposes which we classify as religious and extra-scientific.[40]

The same argument can easily be made for Isaac Newton, generally considered one of the greatest of all Western scientists. Anyone who studies Newton knows—though this fact is often obscured in scientific accounts of his work—that Newton wrote more religious treatises during his life than he did scientific ones. He was something of a biblical fundamentalist, poring over prophetic books like Daniel looking for clues to the unfolding of history. And in his scientific work, he truly believed that he was discerning the mind of God. According to noted historian John Hedley Brooke, for Newton God was not a mere appendage, "a hypothesis to explain what his science could not, a god-of-the-gaps."[41] Newton believed that God's actions were omnipresent in human history and even penetrated his analysis of space. Newton feared that a complete turn to a mechanized universe, one that ran like a machine according to fixed laws, might cripple the idea of divine activity. "Newton insisted that organization could not result from disorganization without the mediation of an intelligent power."[42] With this kind of reasoning, Newton would likely find it difficult to earn tenure today as a physicist in most colleges and universities! It is worth remembering that the strict materialism of contemporary science has not been a constant throughout the history of

---

39. Cited in Wrightsman, "Legitimation of Scientific Belief," 60.
40. Wrightsman, "Legitimation of Scientific Belief," 62.
41. Brooke, *Science and Religion*, 137.
42. Ibid., 144.

science. It is a modern construction and as such can be rethought. Ironically, modern science itself seems to be in the process of deconstructing its own materialist paradigm.

Coming Full Circle

Although the mixing of religion and science that characterized the work of Copernicus and Newton eventually gave way to a more purely materialist science in the eighteenth and nineteenth centuries, twentieth- and now twenty-first-century science may be taking us back to the prematerialist days. As we saw above, quantum theory completely explodes any notion of a determinate universe running according to fixed and ostensibly comprehensible natural laws. And at the level of elementary particles, the material of materialism loses its materiality. It will help to now review another aspect of modern science shaking the foundations of a purely naturalistic understanding of science—the fine-tuning of the universe. As we saw in chapter 1, religion scholars and theologians who uncritically accept the truth of Darwinian evolution feel compelled to give up on traditional religious notions of God as creator. But ironically, science itself may be pointing us back in that direction.

Many contemporary scientists have commented on the fine-tuning of the universe; I will draw my reflections here from the work of Australian cosmologist Paul Davies and his book *The Goldilocks Enigma: Why Is the Universe Just Right for Life?* Davies gives a fascinating account of the many values of physical forces that have to be almost exactly what they are in order to create a universe in which life exists. Many of the values could theoretically be different from what they are. So it seems an incredible coincidence that they just happen to be the values they need to be for life to exist. Were these knobs fine-tuned? And if so, by whom?

Davies begins with a discussion of the expansion rate of the universe. If the universe began in a big bang, as current cosmological theory asserts, the expansion rate would have had some particular value. It just so happens that it has the exact value necessary for the creation of structures like stars, planets, and galaxies. If the universe had expanded a little faster than it did, gravity would not have been able to overcome the effect of the expansion and cause matter to clump together to form structures. The universe would be just a lifeless, amorphous mass of energy. But if the universe had expanded just a little more slowly than it did, gravity would have taken over and caused the nascent universe to collapse back in upon itself. Moreover, cosmologists believe that shortly after the big bang the early universe went through a period of extremely rapid expansion called *inflation*. Such a rapid period of expansion would have stretched space to such a degree that it would have become perfectly smooth—the way that blowing up a balloon smooths out all the wrinkles on the surface. This is a problem, however. Perfectly smooth space would not allow for the formation of structures. But quantum theory fortunately saves the day. Davies writes: "By applying quantum mechanics to inflation, we can predict that some regions of space inflate slightly more or slightly less than others, thus producing a frothy structure superimposed on the overall smoothness of the universe."[43] It is this frothiness, now calculated to be no more than one part in one hundred thousand, that allowed for the congealing of matter into stars and planets, without which there would be no life in the universe.

Even with the perfect expansion rate and the perfect amount of frothiness to permit life, one would expect the elementary particles produced in the big bang to have been evenly divided between particles of matter and particles of antimatter. But had this been the case, we wouldn't be here. The matter and antimatter would have annihilated each other in a flash of pure energy. There must have been slightly more matter produced than antimatter. But why? According to Davies:

> our very existence, not to mention the existence of the visible universe, hinges on the minute degree of symmetry-breaking between matter and antimatter, which in turn depends on how quarks, leptons, and the forces that act between them are amalgamated together in some as yet undetermined grand unification.[44]

Clearly, we don't really know how to account for this slight preference for matter over antimatter. But without it, life would be impossible.

Once we have some left over matter and the universe is expanding at the proper rate, matter can begin to congeal into stars, planets, and galaxies. But the big bang produced only two elements, hydrogen and helium, and the chemistry of life needs many more elements to work with. Where did all the carbon, oxygen, nitrogen, phosphorus, and other heavier elements come from? They were cooked up in the nucleus of stars. Stars

---

43. Davies, *Goldilocks Enigma*, 62.
44. Ibid., 106.

are large nuclear fusion reactors that fuse hydrogen atoms together into helium atoms. But once all the hydrogen is used up, the star will begin fusing helium into heavier elements, and those heavier elements into even heavier ones until the star can no longer balance the inner pull of gravity against the outer push of the heat of fusion. At this point the star will collapse in upon itself and then bounce outward in a supernova explosion that spews all the heavier elements out into space where they can eventually find their way into your body and mine. But there is a catch. It is not easy to fuse carbon, the most important element for life.

Once a star begins to fuse helium, two helium atoms can be fused together to produce beryllium. It is then necessary for a third helium atom to fuse with the beryllium atom to produce carbon. But the nucleus of a beryllium atom fused from two helium atoms is unstable and would be expected to decay before a third helium atom could smash into it to form carbon. So how do stars get past this bottleneck? Enter British astronomer Fred Hoyle. Hoyle theorized that carbon must possess what is called a resonance energy that would prolong the life of the unstable beryllium atom just long enough for a third helium atom to be able to fuse with it. Hoyle even calculated the value of this resonance energy and then convinced skeptical nuclear physicists at the California Institute of Technology in the 1950s to design experiments to detect it. To their great shock, Hoyle was right. This resonance energy existed, and at just the value Hoyle predicted it had to be. The energy of the carbon resonance is determined by the interplay between the strong nuclear force and the electromagnetic force. According to Paul Davies, if the strong nuclear force varied up or down by as little as 1 percent, the carbon resonance would not happen, and the universe would be devoid of carbon and therefore most likely of life. Davies therefore concludes, "The fact that the value of the strong and electromagnetic forces in atomic nuclei are 'just right' for life (like Goldilock's porridge) cries out for explanation."[45]

But what is the explanation? In 1981, Fred Hoyle, despite being an atheist, had this to say:

> From 1953 onward, Fowler and I have been intrigued by the remarkable relation of the 7.65 MeV energy level in the nucleus of 12C to the 7.12 MeV level in 16O. If you wanted to produce carbon and oxygen in roughly equal quantities by stellar nucleosynthesis, these are just the two levels you would have to fix, and your fixing would have to be just about where these levels are

---

45. Ibid., 138.

found to be. Is that another put-up job? Following the above argument, I am inclined to think so. A common sense interpretation of the facts suggests that a superintellect has monkeyed with physics, as well as with chemistry and biology, and that there are no blind forces worth speaking about in nature. The numbers one calculates from the facts seem to me so overwhelming as to put this conclusion almost beyond question.[46]

Clearly Hoyle was deeply impressed by all these apparent coincidences. But are they coincidences? They don't end here.

Once carbon and other elements are fused in the nucleus of stars, those stars have to explode for the elements to get out into interstellar space where they can eventually coalesce into planets and the molecules necessary for life. It turns out that the weak nuclear force is exactly the value it needs to be to promote stellar explosions. If the weak nuclear force varied only slightly from the value it has, stars would not explode and the elements needed for life would remain quarantined in stellar nuclei.

We might also consider the ratio of the masses of elementary particles to each other. The ratio of the mass of a proton to an electron is 1,836.1526675 to 1. The ratio of the mass of a neutron to a proton is 1.00137841870 to 1. These do not look in any way like special numbers. Yet if they varied only slightly from these very ordinary looking values, the chemical reactions necessary for life would likely not occur.

These are just some of the fine-tuning coincidences that render our universe extraordinarily suited to produce life. According to Davies, the Standard Model of particle physics has about twenty undetermined parameters, while cosmology adds about ten more. This makes about thirty fine-tuning knobs that have to be set so precisely that many of them must be set to an accuracy of less than 1 percent for life to be possible.[47] This caused the well-known physicist Freeman Dyson to famously proclaim: "As we look out into the universe and identify the many accidents of physics and astronomy that have worked together to our benefit, it almost seems as if the universe must in some sense have known we were coming."[48] Of course, many hardened materialist scientists will not be comfortable with the idea of a universe fine-tuned and almost

---

46. Hoyle, "Universe," 12.
47. Davies, *Goldilocks Enigma*, 146.
48. Cited in ibid., 223.

preprogrammed to produce life. In order to get around these amazing coincidences, some cosmologists have produced the idea of a multiverse.

If only one universe exists—the one we live in—then the fact that so many undetermined values happen to be exactly what they need to be for life to exist does call out for explanation. But if multiple universes exist, perhaps as many as 10^{500} universes or even an infinite number of them ("infinite number" being an oxymoron if there ever was one!), then the fine-tuning of our universe to be perfectly suited to permit life loses its mystery. If a multiplicity of universes is being created all the time out of random quantum vacuum fluctuations as some cosmologists theorize, then it could well be that each of these multiplicity of universes has a different set of undetermined values. Since most combinations of these values would produce a universe devoid of life, most universes will be lifeless. But if there are an enormous number of such universes in existence, it is bound to happen that at least one of them will have the correct combination of values to produce a life-promoting universe. And that just happens to be ours. In fact, if infinite universes exist, it is guaranteed that at least one will be life-promoting. Actually, it would guarantee that an infinite number of universes exactly like ours must exist, including an infinite number of copies of you and me! With a multiverse, there is no Goldilock's enigma. The fact that our life-promoting universe exists becomes a statistical certainty.

But is the multiverse idea really science? It is impossible for us to ever observe these other universes and know whether they really exist. Some current cosmological theories may predict that a multiplicity of cosmic domains must exist, but how much weight should we place on speculative scenarios that resist empirical confirmation? According to Davies, "The disadvantage of the multiverse theory is that it invokes an overabundance of entities, most of which could never be observed, even in principle. This profligacy strikes many people as an extravagant way to explain bio-friendliness."[49] I am hard-pressed to disagree. The multiverse theory solves the Goldilocks enigma, but at the cost of positing the existence of entities we can never observe. As the well-known atheist turned believing philosopher Anthony Flew put it:

> As I have already mentioned, I did not find the multiverse alternative very helpful. The postulation of multiple universes, I maintained, is a truly desperate alternative. If the existence of

---

49. Ibid., 264.

one universe requires an explanation, multiple universes require a much bigger explanation: the problem is increased by the factor of whatever the total number of universes is. It seems a little like the case of a schoolboy whose teacher doesn't believe his dog ate his homework, so he replaces the first version with the story that a pack of dogs—too many to count—ate his homework.[50]

Despite the evident problems with the multiverse theory, don't expect it to disappear anytime soon. For as Bernard Carr has said so well, "If you don't want God, you'd better have a multiverse."[51]

As far as we know, there is only one universe—ours. It appears to have been fine-tuned for life. How did this happen? No one can say. But the idea of some kind of intelligent force active in the cosmos cannot be discounted just on the grounds that this would not be a scientific explanation, unless we posit the unscientific assumption that science can provide all the answers.

For Harvard astronomer Owen Gingerich:

> Belief in a final cause, a Creator God, gives a coherent understanding of why the universe seems so congenially designed for the existence of intelligent, self-reflective life. Only small changes in numerous physical constants would render the universe uninhabitable. I do not claim the fine-tuning as a proof for the existence of a Creator, only that to me, the universe makes more sense with this understanding, and that is the core of my belief.[52]

I see no grounds on which Gingerich's belief can be discounted. It may not turn out to be true, but we certainly cannot claim to know it is untrue at this stage of our knowledge. It is just as coherent an explanation as any materialist one. It may not be necessary for religion scholars to so easily give up on traditional religious ideas of an intelligent creator, even if the specific ways that creator is described in scriptural and theological sources must be understood as culturally conditioned human interpretation. A mystery still stands at the center of our existence. And as ID advocate Michael Behe has pointed out, "if one allows that a being external to the universe could affect its laws, there is no principled reason to rule out a priori more extensive interaction as well."[53] Is not the mystery that stands

---

50. Flew, *There Is a God*, 137.
51. Cited in Rubenstein, *Worlds without End*, 1.
52. Gingerich, *God's Planet*, 134.
53. Behe, *Edge of Evolution*, 210.

at the center of the fine-tuned universe, the mystery that stands at the center of the origin of life from nonlife in this universe, and the mystery that stands at the center of the evolution and diversity of this life not a fundamentally religious expression? Why do we fear mystery so?

Embracing Mystery

I began this book by observing how University of Chicago biologist James Shapiro has unexpectedly characterized evolution as constituting a continuing mystery whose "fundamental driving forces have not been resolved either in detail or in principle."[54] As I have tried to show throughout this book, Shapiro's view would appear to be an accurate assessment of our current state of knowledge. But most scientists will, of course, balk at such a characterization. Appealing to mystery is seen as abdicating the scientific quest to discover rational, material explanations for all physical phenomena. To call something a mystery is to render it impenetrable and outside the bounds of rational explanation, thus bringing to an abrupt end the human quest for increasing knowledge. Mystery is understood to be anathema to scientific progress.

But where do we get the idea, other than through human hubris, that every aspect of the universe will, given enough time, yield to human rationality? I know of no scientific grounds on which such an idea can be established, and as Noson Yanofsky has demonstrated so well in the previously cited *The Outer Limits of Reason*, there are fundamental limits to what we can know through the use of human reason. This is not to deny the great power of scientific investigation; our world of modern technology stands as a monument to the great power inherent in human reason and ingenuity. It is only to deny science its omnipotent status. Modern science, just like every confessionally specific religious tradition, is a human construct. And as within all human constructs, the actions of power are always bound up with the production of knowledge. Clearly this has been the case with the origin and development of the grand narrative of Darwinian triumph—a scientific narrative that has come to brilliantly serve the guild interests of the community of biologists, but at the expense of truly grappling with the profound nature of the question of the origin and diversity of life.

54. Shapiro, Review of *Darwin's Black Box*.

This was brought home to me in a personal way recently when my daughter came home with her tenth-grade biology textbook. Of course I had to read the chapter on evolution only to find that the chapter begins with the following stunning assertion:

> When Charles Darwin . . . boarded the HMS *Beagle* in 1831, the average person believed that the world was about 6000 years old. Almost everyone, including the young Darwin, believed that animals and plants were unchanging. The concept of gradual change over time—and Darwin's role in this concept—was still years away.[55]

Apparently the pre-Darwinian evolutionary scenarios of Erasmus Darwin, Jean Baptiste Lamarck, and others don't count. Evolution apparently began with Darwin. Such a blatant historical inaccuracy masking as high school science in service to Darwinian mythology! It is obvious why so many people are so easily deceived by Darwin. But this is no longer necessary, especially for scholars of religion.

Of course, this is not to say that religious folks can simply fall back on literal readings of their scriptures and traditions. To undermine the sufficiency of a Darwinian explanation for evolution does not render the Genesis creation account in the Bible historically accurate by default. This is the mistake made by too many creationists and some supporters of ID. While the scientific criticisms of Darwinism made by ID advocates are substantive, and their positive arguments for ID must be seriously engaged, there is still some truth to the criticism that some ID advocates are more interested in using ID to push a conservative Christian agenda than they are in offering a scientific alternative to Darwinism. I once suggested in an email exchange that ID advocates would have natural allies among Muslims, many of whom see secularism as antithetical to their faith. Support for evolution, for example, polls lower in Turkey than it does in the U. S. The Islamophobic pushback I got from this suggestion was telling. For too many ID proponents, maintaining a conservative Christian identity seems more important than the scientific question of the origin of species.

Embracing evolution as a continuing mystery in no way proves the historicity of any confessionally specific religious account of creation, whether Christian, Jewish, Muslim, Hindu, or Native American. All religious language is culturally conditioned and subject to the vagaries of

---

55. Biggs, *Biology*, 418.

human power and privilege. But specific religious constructions aside, embracing the origin and diversity of life as mystery is itself a profoundly religious act in the sense that embracing mystery places us in a position of subordination to something ineffable. And embracing that mystery, rather than trying to dissolve it away either through materialist scientific explanation or through the literal reading of religious texts, may tell us much more about the nature of the reality in which we live than any materialist scientific program ever can.

I realize how difficult this will be for many people. It is difficult for me. I can truly sympathize with an atheistic position, recognizing the utter absurdity of postulating the existence of some guiding intelligence beyond our ability to comprehend when my everyday experience tells me I live in a world of material stuff driven by comprehensible physical laws. But at the same time, I am overcome by the utter absurdity of thinking that this magnificent cosmos and the overwhelming complexity and diversity of life that inhabits Earth are the product of unguided, purely physical, processes. There is absurdity all around, leaving me with little choice but to accept the mystery as mystery and accede to its larger wisdom.

Critics will view my appeal to mystery as just another god-of-the-gaps strategy. Look for the places where materialist explanations seem to fail and then pose a supernatural explanation. Science has been very good at providing materialist explanations for many phenomena that were at one time ascribed to supernatural causation. And as science continues to fill in the gaps, God's role gets smaller and smaller. I know the argument, but I do not think we should be afraid of a god-of-the-gaps strategy. If science is eventually successful at filling in all the gaps by providing a materialist explanation for all phenomena in the universe (or in the vast number of multiverses!), we will just need to have the courage to admit that the atheists were right all along. The universe is a purely material place, and there is no need to pose the existence of God, gods, or any kind of supernatural intelligence. And to do so would be nothing more than to flee from reality. But we are far from that point, and the alternative to a god-of-the-gaps strategy is to turn religious faith into a purely subjective experience, divorcing it from any connection to the physical world and thus rendering it unreal. I would rather make the Pascalian wager that the universe will not ultimately yield to a purely materialist explanation. As we saw in chapter 1, even the great Darwinian paleontologist George Gaylord Simpson admitted that there is likely a great deal in the universe that will never be explained in purely material terms and much that might remain inexplicable on such terms. If this is so, gaps will remain that science can never close, and the profound mystery of our existence will remain a profound mystery. Embracing that mystery as mystery is not an escape from reality then, but a courageous grasping of the way things really are and a humble acknowledgment of the limits of human reason.

Though he is credited with providing a materialist mechanism for the evolutionary process, even Darwin waxed philosophical in the famous closing words of the *Origin of Species*:

> There is grandeur in this view of life, with its several powers, having been originally breathed by the Creator into a few forms or into one; and that, whilst this planet has gone cycling on according to the fixed law of gravity, from so simple a beginning endless forms most beautiful and most wonderful have been, and are being, evolved.[56]

To uncritically accept the view that Darwin solved the problem of the origin of species is to be deceived by Darwin. But to stand with Darwin in recognizing the power and grandeur of the history of life on Earth is to stand squarely in the center of a profound mystery that can hardly fail to awaken the religious sensibilities of anyone who seriously contemplates it.

---

56. Darwin, *On the Origin of Species*, 374.

Bibliography

Allen, Garland E. *Thomas Hunt Morgan: The Man and His Science*. Princeton: Princeton University Press, 1978.
Antonovics, Janis. "The Evolutionary Dys-synthesis: Which Bottles for Which Wine?" *American Naturalist* 129 (1987) 321–31.
Axe, Douglas D. "The Case against a Darwinian Origin of Protein Folds." *Bio-complexity* 2010 (2010) 1–12.
Ayala, Francisco J. "Darwin's Greatest Discovery: Design without Designer." *Proceedings of the National Academy of Sciences, USA* 104 (2007) 8567–73.
———. "On the Virtues and Pitfalls of the Molecular Evolutionary Clock." *Journal of Heredity* 77 (1986) 226–35.
———. "Teleological Explanations in Evolutionary Biology." *Philosophy of Science* 37 (1970) 1–15.
Ayala, Francisco J., and John C. Avise, eds. *Essential Readings in Evolutionary Biology*. Baltimore: Johns Hopkins University Press, 2014.
Baer, Charles C. "Mutation." In *The Princeton Guide to Evolution*, edited by Jonathan B. Losos et al., 315–20. Princeton Reference. Princeton: Princeton University Press, 2014.
Barash, David P. "The Hardest Problem in Science?" Brainstorm: Ideas and Culture. *Chronicle of Higher Education*, October 28, 2011. https://www.chronicle.com/blogs/brainstorm/the-hardest-problem-in-science/40845/.
———. *Natural Selections: Selfish Altruists, Honest Liars, and Other Realities of Evolution*. New York: Bellevue Literary Press, 2008.
Barbour, Ian G. *Religion and Science: Historical and Contemporary Issues*. San Francisco: HarperSanFrancisco, 1997.
Barlow, Nora, ed. *The Autobiography of Charles Darwin, 1809–1892*, by Charles Darwin. With original omissions restored. London: Collins, 1958.
Basener, William F., and John C. Sanford, "The Fundamental Theorem of Natural Selection with Mutations." *Journal of Mathematical Biology* 76 (2017) 1589–1622. doi.org/10.1007/s00285-017-1190x/.
Bateson, Beatrice. *William Bateson, FRS, Naturalist: His Essays and Addresses together with a Short Account of His Life*. Cambridge: Cambridge University Press, 1928.
Bateson, William. *Materials for the Study of Variation*. London: Macmillan, 1894.
Behe, Michael J. *Darwin's Black Box: The Biochemical Challenge to Evolution*. New York: Simon & Schuster, 1996.
———. *The Edge of Evolution: The Search for the Limits of Darwinism*. New York: Free Press, 2008.

Bell, Graham A. C. Review of *Epigenetic Inheritance and Evolution*, by Eva Jablonka and Marion J. Lamb. *Trends in Ecology and Evolution* 11 (1996) 266–67.

———. *Selection: The Mechanism of Evolution*. New York: Chapman & Hall, 1997.

Berman, Marshall. "Intelligent Design Creationism: A Threat to Society—Not Just Biology." *American Biology Teacher* 65 (2003) 646–48. https://doi.org/10.1662/0002-7685(2003)065[0646:IDCATT]2.0.CO;2/.

Biggs, Alton. *Biology*. Student's ed. Glencoe Science. New York: Glencoe/McGraw Hill, 2007.

Blackstone, Neil W. Review of *Darwin's Black Box*, by Michael J. Behe. *Quarterly Review of Biology* 72 (1997) 445–47.

Boesiger, Ernest. "Evolutionary Biology in France at the Time of the Evolutionary Synthesis." In *The Evolutionary Synthesis: Perspectives on the Unification of Biology*, edited by Ernst Mayr and William B. Provine, 309–21. Cambridge: Harvard University Press, 1998 [1980].

Bowler, Peter J. *The Eclipse of Darwinism: Anti-Darwinian Theories in the Decades around 1900*. Baltimore: Johns Hopkins University Press, 1983.

———. *Evolution: The History of an Idea*. 3rd ed., completely rev. and exp. Berkeley: University of California Press, 2003.

Bridges, Bryn A. Review of *Evolution: A View from the 21st Century*, by James A. Shapiro. *DNA Repair* 11 (2012) 99–100.

Bronham, Lindell, et al. "Testing the Cambrian Explosion Hypothesis by Using Molecular Clock Dating Technique." *Proceedings of the National Academy of Sciences, USA* 95 (1998) 12386–89.

Brooke, John Hedley. *Science and Religion: Some Historical Perspectives*. Cambridge History of Science. Cambridge: Cambridge University Press, 1991.

Browne, Janet. *Charles Darwin: The Power of Place*. New York: Knopf, 2002.

Burkhardt, Frederick, et al., eds. *The Correspondence of Charles Darwin*. 26 vols. Cambridge: Cambridge University Press, 1985–2018.

Burkhardt, Richard W., Jr. *The Spirit of System: Lamarck and Evolutionary Biology*. Cambridge: Harvard University Press, 1977.

Cairns, John, et al. "The Origin of Mutants." *Nature* 335 (1988) 142–45.

Cannon, Walter. "Foreword." In *Charles Darwin: The Years of Controversy*, by Peter J. Vorzimmer, xiii–xv. Philadelphia: Temple University Press, 1970.

Carroll, Robert L. "Towards a New Evolutionary Synthesis." *Trends in Ecology and Evolution* 15 (2000) 27–32.

Carson, Hampton. "The Genetics of Speciation at the Diploid Level." *American Naturalist* 109 (1975) 83–92.

Charlesworth, Brian. "Evolution: On the Origins of Novelty and Variation." *Science* 310 (2005) 1619–20.

Charlesworth, Deborah, et al. "The Sources of Adaptive Variation." *Proceedings of the Royal Society B* 284 (2017) 1–12.

Chénais, Benoît, et al. "The Impact of Transposable Elements on Eukaryotic Genomes: From Genome Size Increase to Genetic Adaptation." *Gene* 509 (2012) 7–15.

Chung, Henry, et al. "*Cis*-regulatory Elements in the *Accord* Retrotransposon Result in Tissue-specific Expression of the *Drosophila Melanogaster* Insecticide Gene *Cyp6g1*." *Genetics* 175 (2007) 1071–77.

Cody, Martin L., and Jacob McC. Overton. "Short-Term Evolution of Reduced Dispersal in Island Plant Populations." *Journal of Ecology* 84 (1996) 53–62.

Collins, Francis S. *The Language of God: A Scientist Presents Evidence for Belief.* New York: Free Press, 2006.
Comfort, Nathaniel C. *The Tangled Field: Barbara McClintock's Search for Patterns of Control.* Cambridge: Harvard University Press, 2001.
Conway Morris, Simon. "Evolution: Bringing Molecules into the Fold." *Cell* 100 (2000) 1–11.
———. *Life's Solution: Inevitable Humans in a Lonely Universe.* Cambridge: Cambridge University Press, 2003.
———. *The Runes of Evolution: How the Universe Became Self-Aware.* Conshohocken, PA: Templeton, 2015.
Cook, L. M. "Fryer's New Look at Industrial Melanism in Moths—a Comment." *Linnean* 29 (2013) 10–14.
Cook, L. M., and I. J. Saccheri. "The Peppered Moth and Industrial Melanism: Evolution of a Natural Selection Case Study." *Heredity* 110 (2013) 207–12.
Cornell, John F. Review of *Evolutionary Theory: The Unfinished Synthesis*, by Robert G. B. Reid. *Journal of the History of Biology* 20 (1987) 424–25.
Coyne, Jerry A. "Evolution under Pressure." *Nature* 418 (2002) 19–20.
———. "Not Black and White. Review of *Melanism: Evolution in Action*, by Michael E. N. Majerus. *Nature* 396 (1998) 35–36.
———. *Why Evolution Is True.* New York: Penguin, 2009.
Crick, Francis H. C. "On Protein Synthesis." *Symposium of the Society of Experimental Biology* 12 (1957) 138–63.
Daley, Allison C., et al. "Early Fossil Record of Euarthropoda and the Cambrian Explosion." *Proceedings of the National Academy of Sciences, USA* (2018). doi.org/10.1073/pnas.1719962115.
Darwin, Charles. *The Descent of Man.* The Modern Library of the World's Best Books. New York: Modern Library, 1950.
———. *On the Origin of Species.* The Modern Library of the World's Best Books. New York: Modern Library, 1950.
Darwin, Francis, ed. *The Life and Letters of Charles Darwin.* Vol. 2. Cambridge: Cambridge University Press, 1898. http://darwin-online.org.uk/content/frameset?viewtype=text&itemID=F1452.2&pageseq=1/.
———. *More Letters of Charles Darwin, vol. 2.* London: Murray, 1903.
Davies, Paul. *The Goldilocks Enigma: Why Is the Universe Just Right for Life?* 1st Mariner Books ed. Boston: Houghton Mifflin, 2006.
Dawkins, Richard. *The Blind Watchmaker.* New York: Norton, 1986.
———. *River out of Eden: A Darwinian View of Life.* Science Masters. New York: HarperCollins, 1995.
Deacon, Terrence W. *Incomplete Nature: How Mind Emerged from Matter.* New York: Norton, 2012.
Dembski, William A. "In Defense of Intelligent Design." In *The Oxford Handbook of Religion and Science*, edited by Philip Clayton, 715–31. Oxford Handbooks in Religion and Theology. Oxford: Oxford University Press, 2006.
———. *No Free Lunch: Why Specified Complexity Cannot Be Purchased without Intelligence.* Lanham, MD: Rowman & Littlefield, 2001.
Dennett, Daniel C. *Darwin's Dangerous Idea: Evolution and the Meanings of Life.* New York: Simon & Schuster, 1995.
Denton, Michael. *Evolution: A Theory in Crisis.* Bethesda, MD: Adler & Adler, 1986.

Des Jardins, Julie. *The Madame Curie Complex: The Hidden History of Women in Science*. Women Writing Science. New York: Feminist, 2010.
Dobzhansky, Theodosius. *Evolution, Genetics, and Man*. New York: Wiley, 1955.
———. Foreword. In *Evolution above the Species Level*, by Bernhard Rensch, v. Columbia Biological Series 19. New York: Columbia University Press, 1960.
———. *Genetics and the Origin of Species*. Columbia Biological Series 11. New York: Columbia University Press, 1937.
———. "Nothing in Biology Makes Sense Except in the Light of Evolution." *American Biology Teacher* 35 (1973) 125–29.
———. Review of *The Material Basis of Evolution*, by Richard Goldschmidt. *Science* 92 (1940) 356–58.
Doolittle, Russell F. *The Evolution of Vertebrate Blood Clotting*. Mill Valley, CA: University Science Books, 2013.
Doolittle, Russell F., et al. "Determining Divergence Times of the Major Kingdoms of Living Organisms with a Protein Clock." *Science* 271 (1996) 470–77.
Dronamraju, Krishna, ed. *What I Request from Life: Writings on Science and Life from J. B. S. Haldane*. Oxford: Oxford University Press, 2009.
Eden, Murray. "Inadequacies of Neo-Darwinian Evolution as a Scientific Theory." In *Mathematical Challenges to the Neo-Darwinian Interpretation of Evolution*, edited by Paul S. Moorhead and Martin M. Kaplan, 5–20. Wistar Institute Symposium Monograph 5. Philadelphia: Wistar Institute Press, 1967.
Edis, Taner. "Religion: Accident or Design?" In *The Evolution of Religion: Studies, Theories, and Critiques*, edited by Joseph Bulbulia et al., 387–92. Santa Margarita, CA: Collins Foundation Press, 2008.
Ehrlich, Robert. "What Makes a Theory Testable, or Is Intelligent Design Less Scientific Than String Theory?" *Physics in Perspective* 8 (2006) 83–89.
Eiseley, Loren. *Darwin's Century: Evolution and the Men Who Discovered It*. Doubleday Anchor Books. Garden City, NY: Doubleday, 1958.
Eldredge, Niles, and Stephen Jay Gould. "Punctuated Equilibria: An Alternative to Phyletic Gradualism." In *Models in Paleobiology*, edited by Thomas J. M. Schopf, 82–115. San Francisco: Freeman, Cooper, 1972.
Endler, John. *Natural Selection in the Wild*. Monographs in Population Biology 21. Princeton: Princeton University Press, 1986.
Erwin, Douglas H., and James W. Valentine. *The Cambrian Explosion: The Construction of Animal Biodiversity*. Greenwood Village, CO: Roberts, 2013.
Erwin, Douglas H., et al. "A Comparative Study of Diversification Events: The Early Paleozoic versus the Mesozoic." *Evolution* 41 (1987) 1177–86.
Evolution Institute. "What We Do." *The Evolution Institute* (website). https://evolution-institute.org/about/what-we-do/.
Fawcett, Michael H. "Local and Latitudinal Variation in Predation on an Herbivorous Marine Snail." *Ecology* 65 (1984) 1214–30.
Fisher, Ronald A., Sir. *The Genetical Theory of Natural Selection*. Dover Books on Science. New York: Dover, 1958 [1930].
Flew, Antony, with Roy Abraham Varghese. *There Is a God: How the World's Most Notorious Atheist Changed His Mind*. New York: HarperOne, 2008.
"Flowering Plant Study 'Catches Evolution in the Act.'" *ScienceDaily*, March 18, 2011. https://www.sciencedaily.com/releases/2011/03/110317131034.htm/.
Fryer, Geoffrey. "How Should the History of Industrial Melanism in Moths Be Interpreted?" *The Linnean* 29 (2013) 15–22.

———. "A New—and a Retrospective—Look at Industrial Melanism in Moths." *The Linnean* 28 (2012) 31–36.
Fuller, Steve. *Science vs. Religion? Intelligent Design and the Problem of Evolution*. Cambridge: Polity, 2007.
Futuyma, Douglas J. "Evolution as Fact and Theory." *Bios* 56 (1985) 3–13.
———. "Natural Selection and Adaptation." In *The Princeton Guide to Evolution*, edited by Jonathan B. Losos, 189–91. Princeton Reference. Princeton: Princeton University Press, 2014.
———. *Science on Trial: The Case for Evolution*. New York: Pantheon, 1983.
Gaastra, Wim. Review of *Evolution: A View from the 21st Century*, by James A. Shapiro. *Veterinary Microbiology* 158 (2012) 225.
Gale, Barry G. *Evolution without Evidence: Charles Darwin and "The Origin of Species."* Albuquerque: University of New Mexico Press, 1982.
Ghiselin, Michael T. *The Triumph of the Darwinian Method*. Berkeley: University of California Press, 1969.
Gingerich, Owen. *God's Planet*. Cambridge: Harvard University Press, 2014.
Goldschmidt, Richard B. "Evolution, as Viewed by One Geneticist." *American Scientist* 40 (1952) 84–125.
———. *The Material Basis of Evolution*. Silliman Milestones in Science. New Haven: Yale University Press, 1982 [1940].
Gould, Fred. "Rapid Host Range Evolution in a Population of the Phytophagous Mite Tetranychus Urticae Koch." *Evolution* 33 (1979) 791–802.
Gould, Stephen Jay. "G. G. Simpson, Paleontology, and the Modern Synthesis." In *The Evolutionary Synthesis: Perspectives on the Unification of Biology*, edited by Ernst Mayr and William B. Provine, 153–72. With a new preface by Ernst Mayr. Cambridge: Harvard University Press, 1998[1980].
———. "Is a New and General Theory of Evolution Emerging?" *Paleobiology* 6 (1980) 119–30.
———. "Nonoverlapping Magisteria." This View of Life. *Natural History* 106/2 (1997) 16–22.
———. *The Structure of Evolutionary Theory*. Cambridge: Belknap, 2002.
———. "The Uses of Heresy: An Introduction to Richard Goldschmidt's *The Material Basis of Evolution*." In *The Material Basis of Evolution*, by Richard B. Goldschmidt, xiii–xlii. Silliman Milestones in Science. New Haven: Yale University Press, 1982 [1940].
———. *Wonderful Life: The Burgess Shale and the Nature of History*. New York: Norton, 1989.
Gould, Stephen Jay, and Niles Eldredge. "Punctuated Equilibria: The Tempo and Mode of Evolution Reconsidered." *Paleobiology* 3 (1977) 115–51.
Greene J. C. "The History of Ideas Revisited." *Review de Synthèse* 4 (1986) 201–27.
Grose, Jonathan. Review of *Evolution in Four Dimensions*, by Eva Jablonka and Marion J. Lamb. *British Journal for the Philosophy of Science* 60 (2009) 667–72.
Haldane, J. B. S. *The Causes of Evolution*. Ithaca: Cornell University Press, 1966 [1932].
Hall, Barry G. "Adaptive Evolution that Requires Multiple Spontaneous Mutations: Mutations Involving Base Substitutions." *Proceedings of the National Academy of Sciences, USA* 88 (1991) 5882–86.
———. "Adaptive Evolution that Requires Multiple Spontaneous Mutations: Mutations Involving an Insertion Sequence." *Genetics* 120 (1988) 887–97.

Hamburger, Viktor. "Embryology." In *The Evolutionary Synthesis: Perspectives on the Unification of Biology*, edited by Ernst Mayr and William B. Provine, 97–112. With a new preface by Ernst Mayr. Cambridge: Harvard University Press, 1998[1980].

Hamer, Dean. *The God Gene: How Faith Is Hardwired into Our Genes*. New York: Anchor, 2005.

Harrison, Peter. *The Territories of Science and Religion*. Chicago: University of Chicago Press, 2015.

Haught, John F. *Deeper Than Darwin: The Prospect for Religion in the Age of Evolution*. Boulder: Westview, 2003.

———. *God after Darwin: A Theology of Evolution*. Boulder: Westview, 2000.

Heisenberg, Werner. *Physics and Beyond: Encounters and Conversations*. Translated by Arnold J. Pomerans. World Perspectives 42. New York: Harper & Row, 1971.

Hennessey, Elizabeth. "Mythologizing Darwin's Islands." In *Darwin, Darwinism, and Conservation in the Galápagos Islands: The Legacy of Darwin and Its New Applications*, edited by Diego Quiroga and Ana Sevilla, 65–90. Social and Ecological Interactions in the Galapagos Islands. Cham, Switzerland: Springer, 2017.

Himmelfarb, Gertrude. *Darwin and the Darwinian Revolution*. New York: Norton, 1959.

Ho, Mae-wan. "Why Lamarck Won't Go Away." *Annals of Human Genetics* 60 (1996) 81–84.

Hoekstra, H. E., et al. "Strength and Tempo of Directional Selection in the Wild." *Proceedings of the National Academy of Sciences* 98 (2001) 9157–60.

Hopwood, Nick. *Haeckel's Embryos: Images, Evolution, and Fraud*. Chicago: University of Chicago Press, 2015.

Hoyle, Fred. "The Universe: Past and Present Reflections." *Engineering & Science* (Nov. 1981) 8–12. http://calteches.library.caltech.edu/527/2/Hoyle.pdf/.

Huxley, Julian. *Evolution: The Modern Synthesis*. With a new foreword by Massimo Pigliucci and Gerd B. Müller. Definitive ed. Cambridge: MIT Press, 2010.

Huxley, Thomas Henry. *Darwiniana Essays*. New York: Appleton, 1896 [1893].

Jablonka, Eva, and Marion J. Lamb. *Epigenetic Inheritance and Evolution: The Lamarckian Dimension*. New York: Oxford University Press, 1995.

———. *Evolution in Four Dimensions: Genetic, Epigenetic, Behavioral and Symbolic Variation in the History of Life*. Life and Mind. Cambridge: MIT Press, 2005.

Jacob, François. "Evolution and Tinkering." *Science* 196 (1977) 1161–66.

———. *The Possible and the Actual*. The Jessie and John Danz Lectures. Seattle: University of Washington Press, 1982.

Jensen, Derrick. *The Myth of Human Supremacy*. New York: Seven Stories, 2016.

Johnson, Norman A. "Is Evolution 'Only a Theory'? Scientific Methodologies and Evolutionary Biology." In *Scientists Confront Intelligent Design and Creationism*, edited by Andrew J. Petto and Laurie R. Godfrey, 339–60. New York: Norton, 2007.

Judson, Horace Freeland. *The Eighth Day of Creation: Makers of the Revolution in Biology*. Cold Spring Harbor, NY: Cold Spring Harbor Laboratory Press, 1996.

Kadanoff, Leo P. "Intelligent Design and Complexity Research." *Journal of Statistical Physics* 110 (2003) 451–55.

Keller, Evelyn Fox. *A Feeling for the Organism: The Life and Work of Barbara McClintock*. San Francisco: Freeman, 1983.

Kelly, Edward F., et al., eds. *Beyond Physicalism: Toward Reconciliation of Science and Spirituality*. Lanham, MD: Rowman & Littlefield, 2015.
Kemp, T. S. *The Origin of Higher Taxa: Palaeobiological, Developmental, and Ecological Perspectives*. Oxford: Oxford University Press, 2016.
Kingsolver, J. G., et al. "The Strength of Phenotypic Selection in Natural Populations." *American Naturalist* 157 (2001) 245–61.
Kirschner, Marc W., and John C. Gerhart. *The Plausibility of Life: Resolving Darwin's Dilemma*. New Haven: Yale University Press, 2005.
Knoll, Andrew H., and Sean B. Carroll. "Early Animal Evolution: Emerging Views from Comparative Biology and Geology." *Science* 284 (1999) 2129–37.
Kröger, Ronald H. H., and Oliver Biehlmaier. "Space-Saving Advantage of an Inverted Retina." *Vision Research* 49 (2009) 2318–21.
Kuhn, Thomas S. *The Structure of Scientific Revolutions*. 2nd ed. Chicago: University of Chicago Press, 1970 [1962].
Larson, Edward J. *Evolution: The Remarkable History of a Scientific Theory*. Modern Library Chronicles 17. New York: Modern Library, 2004.
Lenski, Richard. "Are Some Mutations Directed?" *Trends in Ecology & Evolution* 4 (1989) 148–50.
Lenski, Richard, and John E. Mittler. "The Directed Mutation Controversy and Neo-Darwinism." *Science* 259 (1993) 188–94.
Lenski, Richard, et al. "Mutation and Selection in Bacterial Populations: Alternatives to the Hypothesis of Directed Mutation." *Proceedings of the National Academy of Sciences, USA* 86 (1989) 2775–78.
Lessl, Thomas M. *Rhetorical Darwinism: Religion, Evolution, and the Scientific Identity*. Studies in Rhetoric and Religion 11. Waco: Baylor University Press, 2012.
Levinton, Jeffrey S. "The Big Bang of Animal Evolution." *Scientific American* 84 (1992) 84–93.
Lewontin, Richard C. "Billions and Billions of Demons." *New York Review of Books* (January 9, 1997). https://www.nybooks.com/articles/1997/01/09/billions-and-billions-of-demons/.
Løvtrup, Søren. *Darwinism: The Refutation of a Myth*. London: Croom Helm, 1987.
Luria S. E., and Max Delbrück. "Mutations of Bacteria from Virus Sensitivity to Virus Resistance." *Genetics* 28 (1943) 491–511.
Lyell, Charles. *The Antiquity of Man*. Everyman's Library. London: Dent, 1927 [1863].
Lyons, Sherrie L. *Thomas Henry Huxley: The Evolution of a Scientist*. Amherst, NY: Prometheus, 1999.
MacPhee, Donald. "Directed Evolution Reconsidered." *American Scientist* 81 (1993) 554–61.
Majerus, Michael E. N. *Melanism: Evolution in Action*. Oxford: Oxford University Press, 1998.
———. "The Peppered Moth: Decline of a Darwinian Disciple." In *Insect Evolutionary Ecology: Proceedings of the Royal Entomological Society's 22nd Symposium*, edited by M. D. E. Fellowes et al., 371–96. Wellingford, UK: CABI, 2005.
Marchant, Jo. *Cure: A Journey into the Science of Mind over Body*. New York: Crown, 2016.
Martens, Koen. Review of *Epigenetic Inheritance and Evolution*, by Eva Jablonka and Marion J. Lamb. *Biological Journal of the Linnean Society* 73 (2001) 341–44.

Matthews, R. "Scientists Pick Holes in Darwin Moth Theory." *Daily Telegraph*, March 18, 1999.

Matzke, Nichols J. Review of *The Edge of Evolution*, by Michael J. Behe. *Trends in Ecology and Evolution* 22 (2007) 566–67.

Mayr, Ernst. *Animal Species and Evolution*. Cambridge: Belknap, 1965.

———. *Evolution and the Diversity of Life: Selected Essays*. Cambridge: Belknap, 1976.

———. "Evolution and God." *Nature* 248 (1974) 285–86.

———. "How I Became a Darwinian." In *The Evolutionary Synthesis: Perspectives on the Unification of Biology*, edited by Ernst Mayr and William B. Provine, 413–23. With a new preface by Ernst Mayr. Cambridge: Harvard University Press, 1998 [1980].

———. *One Long Argument: Charles Darwin and the Genesis of Modern Evolutionary Thought*. Questions of Science. Cambridge: Harvard University Press, 1991.

———. "Preface, 1998." In *The Evolutionary Synthesis: Perspectives on the Unification of Biology*, edited by Ernst Mayr and William B. Provine, ix–xiv. With a new preface by Ernst Mayr. Cambridge: Harvard University Press, 1998 [1980].

———. "Preface to the Original Edition." In *The Evolutionary Synthesis: Perspectives on the Unification of Biology*, edited by Ernst Mayr and William B. Provine, xv–xvi. With a new preface by Ernst Mayr. Cambridge: Harvard University Press, 1998 [1980].

———. "Prologue: Some Thoughts on the History of the Evolutionary Synthesis." In *The Evolutionary Synthesis: Perspectives on the Unification of Biology*, edited by Ernst Mayr and William B. Provine, 1–48. With a new preface by Ernst Mayr. Cambridge: Harvard University Press, 1998 [1980].

———. Review of *Evolutionary Theory: The Unfinished Synthesis*, by Robert G. B. Reid. *Isis* 77 (1986) 358–59.

———. "The Role of Systematics in the Evolutionary Synthesis." In *The Evolutionary Synthesis: Perspectives on the Unification of Biology*, edited by Ernst Mayr and William B. Provine, 123–36. With a new preface by Ernst Mayr. Cambridge: Harvard University Press, 1998 [1980].

———. *Systematics and the Origin of Species*. Dover Books on the Biological Sciences. New York: Dover, 1964 [1942].

———. *Toward a New Philosophy of Biology*. Cambridge: Harvard University Press, 1988.

———. "The Triumph of Evolutionary Synthesis." *Times Literary Supplement* (Nov. 2, 1984) 1261–62.

———. *What Evolution Is*. New York: Basic Books, 2001.

———. "Where Are We?" *Cold Spring Harbor Symposium on Quantitative Biology* 24 (1959) 1–14.

Mayr, Ernst, and William B. Provine, eds. *The Evolutionary Synthesis: Perspectives on the Unification of Biology*. With a new preface by Ernst Mayr. Cambridge: Harvard University Press, 1998 [1980].

Mazur, Suzan. *The Altenberg 16: An Exposé of the Evolution Industry*. Berkeley: North Atlantic, 2009.

McClintock, Barbara. "The Significance of Responses of the Genome to Challenge." *Science* 226 (1984) 792–801.

Medawar, Peter B. Review of *The Phenomenon of Man*, by Pierre Teilhard de Chardin. Critical Notice. *Mind* 70 (1961) 99–106.

Meyer, Stephen C. *Darwin's Doubt: The Explosive Origin of Animal Life and the Case for Intelligent Design*. New York: HarperOne, 2013.
Miller, Kenneth R. *Finding Darwin's God: A Scientist's Search for Common Ground between God and Evolution*. New York: Cliff Street, 1999.
Mivart, St. George Jackson, FRS. *On the Genesis of Species*. New York: Appleton, 1871.
Morgan, Thomas Hunt. *Evolution and Genetics*. Princeton: Princeton University Press, 1925.
———. "For Darwin." *Popular Science Monthly* 74 (1909) 367–80.
Müller, Gerd B., and Stuart A. Newman, eds. *Origination of Organismal Form: Beyond the Gene in Developmental and Evolutionary Biology*. Cambridge: MIT Press, 2003.
Nagel, Thomas. *Mind and Cosmos: Why the Materialist Neo-Darwinian Conception of Nature Is Almost Certainly False*. New York: Oxford University Press, 2012.
———. "Public Education and Intelligent Design." *Philosophy and Public Affairs* 36 (2008) 187–205.
Newman, Stuart A. "The Demise of the Gene." *Capitalism Nature Socialism* 24 (2013) 62–72.
Nilsson, Dan-E. "Eye Evolution: Its Functional Basis." *Visual Neuroscience* 30 (2013) 5–20.
Nilsson, Dan-E., and Susanne Pelger. "A Pessimistic Estimate of the Time Required for an Eye to Evolve." *Proceedings of the Royal Society of London B* 256 (1994) 53–58.
Noble, Denis. *Dance to the Tune of Life: Biological Relativity*. Cambridge: Cambridge University Press, 2017.
Norenzayan, Ara. *Big Gods: How Religion Transformed Cooperation and Conflict*. Princeton: Princeton University Press, 2013.
Orr, H. Allen. "Response to Berlinski." In *Uncommon Dissent: Intellectuals Who Find Darwinism Unconvincing*, edited by William A. Dembski, 283–86. Wilmington, DE: ISI, 2004.
Palevitz, Barry A. "Intelligent Design Creationism: None of Your Business? Think Again." *Evolution* 56 (2002) 1718–20.
Parisod, Christian, et al. "Impact of Transposable Elements on the Organization and Function of Allopolyploid Genomes." *New Phytologist* 186 (2010) 37–45.
Park, Lisa E. "It's Not about Evolution: The Debate about Intelligent Design." Spotlight. *Palaios* 21 (2006) 111–13.
Pennisi, Elizabeth. "Modernizing the Modern Synthesis." *Science* 321 (2008) 196–97.
Pennock, Robert T. "The Pre-Modern Sins of Intelligent Design." In *The Oxford Handbook of Religion and Science*, edited by Philip Clayton, 732–48. Oxford Handbooks in Religion and Theology. Oxford: Oxford University Press, 2006.
Penny, David. Review of *Evolution: A View from the 21st Century*, by James A. Shapiro. *Systematic Biology* 61 (2012) 709–10
Perlman, Robert L. Review of *The Plausibility of Life*, by Marc W. Kirschner and John C. Gerhart. *Perspectives in Biology and Medicine* 50 (2007) 314–16.
Peterson, Michael, and Michael Ruse. *Science, Evolution, and Religion: A Debate about Atheism and Theism*. New York: Oxford University Press, 2017.
Pigliucci, Massimo. "The Proper Role of Population Genetics in Modern Evolutionary Theory." *Biological Theory* 3 (2008) 316–24.
Pigliucci, Massimo, and Gerd Müller, eds. *Evolution: The Extended Synthesis*. Cambridge: MIT Press, 2010.

Pigliucci, Massimo, and Jonathan Kaplan. *Making Sense of Evolution: The Conceptual Foundations of Evolutionary Biology.* Chicago: University of Chicago Press, 2006.

Provine, William B. "Evolution and the Foundation of Ethics." *Marine Biological Laboratories Science* 3 (1988) 25–29.

———. *The Origins of Theoretical Population Genetics.* The Chicago History of Science and Medicine. Chicago: University of Chicago Press, 1971.

———. *Sewall Wright and Evolutionary Biology.* Science and Its Conceptual Foundations. Chicago: University of Chicago Press, 1986.

Ramón y Cajal, Santiago. *Recollections of My Life.* 2 vols. Philadelphia: American Philosophical Society, 1937.

Reid, Robert G. B. *Biological Emergences: Evolution by Natural Experiment.* The Vienna Series in Theoretical Biology. Cambridge: MIT Press, 2007.

Rensch, Bernhard. *Evolution above the Species Level.* Columbia Biological Series 19. New York: Columbia University Press, 1960.

Richards, O. W., and G. C. Robson. "The Species Problem and Evolution." *Nature* 117 (1926) 382–8.

Robson, G. C. *The Species Problem.* Biological Monographs and Manuals 8. Edinburgh: Oliver & Boyd, 1928.

Rose, Christopher S. Review of *Biological Emergences*, by Robert G. B. Reid. *Integrative and Comparative Biology* 48 (2008) 871–73.

Rubenstein, Mary-Jane. *Worlds without End: The Many Lives of the Multiverse.* New York: Columbia University Press, 2014.

Ruse, Michael. "Dobzhansky and the Problem of Progress." In *The Evolution of Theodosius Dobzhansky*, edited by Mark B. Adams, 233–46. Princeton: Princeton University Press, 1994.

———. *The Evolution-Creation Struggle.* Cambridge: Harvard University Press, 2005.

Salisbury, Frank B. "Natural Selection and the Complexity of the Gene." *Nature* 224 (1969) 342–43.

Salvini-Plawen, L. V., and Ernst Mayr. "On the Evolution of Photoreceptors and Eyes." *Evolutionary Biology*, vol. 10, edited by Max K. Hecht et al., 207–63. New York: Plenum, 1977.

Sargent, Theodore D., et al. "The 'Classical' Explanation of Industrial Melanism." *Evolutionary Biology* 30 (1998) 299–322.

Schaab, Gloria L. "Evolutionary Theory and Theology: A Mutually Illuminative Dialogue." *Zygon* 43 (2008) 9–19.

Schlichting, Carl D. Review of *The Plausibility of Life*, by Marc W. Kirschner and John C. Gerhart. *Quarterly Review of Biology* 81 (2006) 169–70.

Schluter, Dolph. "Ecology and the Origin of Species." *Trends in Ecology and Evolution* 16 (2001) 372–80.

Schwartz, Jeffrey H. *Sudden Origins: Fossils, Genes, and the Emergence of Species.* New York: Wiley, 1999.

Shallit, Jeffrey. Review of *No Free Lunch*, by William A. Dembski. *Biosystems* 66 (2002) 93–99.

Shapiro, James A. *Evolution: A View from the 21st Century.* Safari Books Online. Upper Saddle River, NJ: FT Press Science, 2011.

———. Review of *Darwin's Black Box*, by Michael J. Behe. *National Review*, September 16, 1996.

Shaw, George Bernard. *Back to Methuselah: A Metabiological Pentateuch*. New York: Brentano, 1921.
Simpson, George Gaylord. *The Major Features of Evolution*. New York: Columbia University Press, 1953.
———. *The Meaning of Evolution: A Study of the History of Life and Its Significance*. Rev. ed. New Haven: Yale University Press, 1967.
———. *Tempo and Mode in Evolution*. Columbia Biological Series 15. New York: Columbia University Press, 1944.
Smocovitis, Vassiliki Betty. *Unifying Biology: The Evolutionary Synthesis and Evolutionary Biology*. Princeton: Princeton University Press, 1996.
Soltis, Douglas E., et al. "What We Still Don't Know about Polyploidy." *Taxon* 59 (2010) 1387–1403.
Spitzer, Jan. "Emergence of Life on Earth: A Physicochemical Jigsaw Puzzle." *Journal of Molecular Evolution* 84 (2017) 1–7.
Stearns, Stephen C. "Natural Selection, Adaptation, and Fitness: Overview." In *The Princeton Guide to Evolution*, edited by Jonathan B. Losos, 193–99. Princeton Reference. Princeton: Princeton University Press, 2014.
Stebbins, G. Ledyard. *Variation and Evolution in Plants*. Columbia Biological Series 16. New York: Columbia University Press, 1950.
Sulloway, Frank J. "Darwin and His Finches: The Evolution of a Legend." *Journal of the History of Biology* 15 (1982) 1–53.
Symonds, Neville. "Anticipatory Mutagenesis?" *Nature* 337 (1989) 119–20.
———. "A Fitter Theory of Evolution?" *New Scientist* 131 (1991) 30–34. https://www.newscientist.com/article/mg13117874-200/.
Taylor, Charles E. "Dobzhansky, Artificial Life, and the 'Larger Questions' of Evolution." In *The Evolution of Theodosius Dobzhansky*, edited by Mark B. Adams, 163–76. Princeton: Princeton University Press, 1994.
Thomas, Brian. "Did Flower Study Catch Evolution in the Act?" *Institute for Creation Research*. www.icr.org/article/did-flower-study-catch-evolution-act/.
Thomson, K. S. "Macroevolution: The Morphological Problem." *American Zoologist* 32 (1992) 106–12.
Tian, Pengfei and Robert B. Best. "How Many Protein Sequences Fold to a Given Structure? A Coevolutionary Analysis." *Biophysical Journal* 113 (2017) 1719–30.
Understanding Evolution. https://evolution.berkeley.edu/.
Valentine, James, et al. "Fossils, Molecules, and Embryos: New Perspectives on the Cambrian Explosion. *Development* 126 (1999) 851–59.
Vopalensky, Pavel, et al. "Molecular Analysis of the Amphioxus Frontal Eye Unravels the Evolutionary Origin of the Retina and Pigment Cells of the Vertebrate Eye." *Proceedings of the National Academy of Sciences, USA* 109 (2012) 15383–88.
Vorzimmer, Peter J. *Charles Darwin: The Years of Controversy; "The Origin of Species" and Its Critics*. Philadelphia: Temple University Press, 1970.
Wallace, Alfred Russel. "On the Tendency of Varieties to Depart Indefinitely from the Original Type." *Journal of the Proceedings of the Linnean Society (Zoology)* 3 (1858) 53–62.
Watling, Jennifer, et al. "Impact of Pre-Columbian 'Geoglyph' Builders in Amazonian Forests." *Proceedings of the National Academy of Sciences, USA* 114 (2017) 1868–73.
Watson, James D. *The Double Helix*. New York: Atheneum, 1968.

Watson, J. D., and F. H. C. Crick, "Molecular Structure of Nucleic Acids." *Nature* 171 (1953) 737–38.
Weismann, August. "The Selection Theory." In *Darwin and Modern Science*, edited by A. C. Seward, 18–65. Cambridge: Cambridge University Press, 1909.
Wells, Jonathan. *Icons of Evolution: Science or Myth?* Washington DC.: Regnery, 2000.
Whitfield, John. "Postmodern Evolution?" *Nature* 455 (2008) 281–84.
Wilkins, Adam S. Review of *Evolution: A View from the 21st Century*, by James A. Shapiro. *Genome Biology and Evolution* 4 (2012) 423–26.
Wilkins, John S. "Could God Create Darwinian Accidents?" *Zygon* 47 (2012) 30–42.
Wilson, Edward O. *On Human Nature*. Cambridge: Harvard University Press, 1978.
Woese, C. R. "On the Evolution of the Genetic Code." *Proceedings of the National Academy of Sciences, USA* 54 (1965) 1546–52.
Wolf, Stewart. Review of *Darwin's Black Box*, by Michael J. Behe. *Integrative Physiological and Behavioral Science* 32 (1997) 417–18.
Woollacott, Marjorie Hines. *Infinite Awareness: The Awakening of a Scientific Mind*. Lanham, MD: Rowman & Littlefield, 2015.
Wray, Gregory A. "Evolution and Development." In *Evolution: The First Four Billion Years*, edited by Michael Ruse and Joseph Travis, 208–36. Cambridge: Belknap, 2009.
Wray, Gregory A., et al. "Molecular Evidence for Deep Precambrian Divergences among Metazoan Phyla." *Science* 274 (1996) 568–81.
Wright, Sewall. "Gene and Organism." *American Naturalist* 87 (1953) 5–18.
———. Review of *The Material Basis of Evolution*, by Richard Goldschmidt. *Scientific Monthly* 53 (1941) 165–70.
Wrightsman, Bruce. "The Legitimation of Scientific Belief: Theory Justification by Copernicus." In *Scientific Discovery: Case Studies*, edited by Thomas Nickles, 51–66. Boston Studies in the Philosophy of Science 60. Dordrecht: Reidel, 1980.
Yanofsky, Noson S. *The Outer Limits of Reason: What Science, Mathematics, and Logic Cannot Tell Us*. Cambridge: MIT Press, 2016.
Zygon: Journal of Religion and Science. "Who We Are: *Zygon*'s Statement of Perspective." http://www.zygonjournal.org/about.html/.

Index

Agassiz, Alexander, 41
allopatric speciation, 142–43, 200–201, 203; see geographic speciation
The Altenberg 16, 30, 214–16
Anderson, Edger, 87
Antonovics, Janis, 90, 198
Archaeopteryx, 48
Avise, John C., 86, 88, 236
Axe, Douglas, 221, 225, 227
Ayala, Francisco J.
   design without designer, 193
   on directed mutation, 176–77
   on molecular clock dating, 151
   on non-overlapping magisteria, 236
   on population genetics, 86, 88
   and teleological analogies for natural selection, 29, 189
   on teleology, 190

Baer, Charles, 180
Bailey, George, 52
Barash, David P., 47, 49, 192–93, 248
Barbour, Ian, 5–8
Barton, Nicholas, H., 181
Basener, William, 80
Bateson, William, 74, 135, 183
   on coordinated mutation, 72, 106
   Ernst Mayr's criticism of, 188
   influence on William Castle, 77
   and Mendelian genetics, 14, 25, 164
   Ronald Fisher's criticism of, 78
   and the study of variation, 71–73
"beanbag genetics," 85
Behe, Michael J., xii, 1, 23, 259
   and irreducible complexity, 223
   James Shapiro's review of, 1
   marginalization of, 229
   Neil Blackstone's criticism of, 21
   Nicholas Matzke's criticism of, 21
   professional credentials of, 221
   Stuart Wolf's review of, 21
Bell, Graham, 30, 96, 109, 129–30, 192
   on directed mutation, 180
   on epigenetics, 207
   on natural selection, 110–14
Bergson, Henri, 67
Berlinski, David, 221
Berman, Marshall, 20
Best, Robert, 227
Biehlmaier, Oliver, 63
big bang, 241, 255
biological relativity, 209–10
Blackstone, Neil, 21
Bohr, Niels, 244
Bowler, Peter, 66, 69, 78
Brenner, Sydney, 206
Bridges, Bryn, 185
Bronham, Lindell, 154
Brooke, John Hedley, 253
Bruecke, E. von, 186n73
Buggs, Richard, 162–63
Burgess Shale fossils, 52–53, 147

Burke, Edmund, 20
Burnet, Sir Macfarlane, 168
Bush, Guy, 144

Cairns, John, 28, 173–79, 180–81, 217
Cambrian explosion, 30, 51–52, 130, 133, 146–59, 203
Cannon, Walter, 41
Capra, Frank, 52
Carr, Bernard, 259
Carroll, Robert, 203
Carroll, Sean B., 148
Carson, Hampton, 145
Castle, William, 77
catastrophism (in geology), 35, 37, 228
Central Dogma of molecular biology
    criticisms of, 169–71, 173, 182, 185
    and Darwinian evolution, 166–67
    definition of, 27–28, 166
    and directed mutation, 179
    and the extended evolutionary synthesis, 216
    lack of evidence for, 167–68
Chambers, Robert, 38
Chargaff, Erwin, 166
Charlesworth, Brian, 181, 218
Charlesworth, Deborah, 181
Chengjiang fauna, 147
Chung, Henry, 186
Cody, Martin L., 115
Collins, Francis, 238, 240–42
combinatorial inflation, 113, 223–27
Comfort, Nathan, 170–71
convergent evolution (or evolutionary convergence), 30, 111, 138
    and analogies to evolution, 191–92
    Ernst Mayr on, 55
    and evolutionary constraint, 53, 195
    and eye evolution, 56
    George Ledyard Stebbins on, 55–56
    Simon Conway Morris on, 53–54
    St. George Jackson Mivart on, 50–51
Conway Morris, Simon, 51–55, 67, 111
Cook, Laurence, 127
Copernicus, Nicholas, 251–54
Cope's Rule, 136–37
Cornell, John F., 213–14
correlated progression, 156–58
Coyne, Jerry, 127, 243
    on the extended evolutionary synthesis, 215
    on eye evolution, 58–59
    on the fossil record, 154
    on industrial melanism, 119–23
Crick, Francis, xiii, 27–28, 90, 164, 182
    atheism of, 168–69
    on the Central Dogma of molecular biology, 166–68
    on the structure of DNA, 165–66
Crozier, W. J., 89

Daley, Allison, 155–56
Darwin, Charles, xii–xv, 5, 8, 11, 16–17, 28, 30, 47n30, 50–51, 55, 69, 75, 78–80, 92–93, 97–98, 103, 111, 118, 138, 142, 158, 164, 212, 214, 216, 228
    and artificial selection, 29, 187, 229
    assessment of, 39–49
    biographical sketch of, 33–39, 64
    and the Cambrian explosion, 146–50
    criticisms of, 10, 13–14, 24–26, 49, 65–67, 71, 74, 86
    and the directed mutation controversy, 176–79
    and eye evolution, 7, 56–58, 60, 64

and the fossil record, 52, 106
  132–35, 139, 146–49, 199–
  200, 203–4
as foundation for atheism, 220
hagiographical treatment of,
  1–2, 10, 36n6, 193, 261
ignorance of Mendelian
  genetics, 70
and Jean Baptiste Lamarck,
  36–37
and natural theology, 18–19
and the *Origin of Species*, 32, 263
population thinking of, 14–15
responses to criticism, 49n40
and Thomas Henry Huxley,
  47–49, 163, 183
Darwin, Erasmus (brother), 34, 46
Darwin, Erasmus (grandfather),
  36, 261
Darwin, Robert, 33–34
Darwin, Susan, 34
Darwin, Susannah Wedgwood, 33
Davies, Paul, 254–58
Dawkins, Richard, xii–xiii, 2, 11,
  62–63, 220, 242
Dead Sea Scrolls, 153
Delbrück, Max, 173–74
Dembski, William A., 21, 221, 246
Dennett, Daniel, xiii, 2, 4, 13, 33,
  102–3, 247
Denton, Michael, 18, 20, 23, 42,
  80, 92
De Vries, Hugo, 70, 72–74, 77, 183
  and Mendelian genetics, 14, 25,
    70–71
  on *mutationstheorie*, 71
  as a saltationist, 14, 25, 164
dialectical materialism, 84
directed mutation, 30, 115, 169, 217
  in bacterial experiments, 173–74
  as a Darwinian heresy, 28,
    176–81, 186
  and epigenetics, 199, 206–8
  and "natural experiments," 212
Discovery Institute, 230
Dobzhansky, Theodosius, xii–xiii,
  10, 90, 104, 109, 128, 144,
  164, 238

biographical sketch of, 97
as a creationist, 103
on directed mutation, 180–81
and the grand narrative of
  Darwinian triumph, 15–16,
  26–27, 87, 94, 96, 101
on natural selection, 99–100
on Pierre Teilhard de Chardin,
  102–3
on polyploidy, 159
on population size, 99
on Richard Goldschmidt, 27,
  143
on Sewall Wright, 87
on the "species problem," 97
as a theistic evolutionist, 242–44
on variation, 98
Doolittle, Russell F., 223–24
Dyson, Freeman, 257

Easter Island, 228
eclipse of Darwinism, 14, 25, 66, 69,
  77, 118
Eden, Murray, 113
Ediacaran fauna, 150
Edis, Taner, 5
Ehrlich, Robert, 21
Einstein, Albert, 26, 79, 209, 244,
  250
Eiseley, Loren, 43
élan vital, 67
Eldredge, Niles, 29, 199–203, 212
Emu Bay Shale, 152
Endler, John, 30, 96, 109–10, 114–
  15, 128–29
epigenetic inheritance, 30, 62, 177,
  199, 204–7, 209, 212, 216
epistasis, 77
Erwin, Douglas H., 148–49, 151–52
eugenics, 66, 78, 80
Evolution Institute, 13

facilitated variation, 216–18
Fawcett, Michael H., 114
Fisher, Ronald, 84–85, 87–88, 90,
  99, 140, 185
  contemporary criticisms of, 80

Fisher, Ronald (*continued*)
  differences with Sewall Wright, 81–83, 85
  and eugenics, 78
  and the Fundamental Theorem of Natural Selection, 79–80, 91
  and population genetics, 15, 26, 77–79
  and the quantification of biology, 79–80, 88–89, 93
Fitzroy, Robert, 35
Flew, Anthony, 258
Franklin, Rosalind, 166, 166n2
Frazzetta, T. H., 144
Fryer, Geoffrey, 126–27
Fuller, Steve, 90
Futuyma, Douglas, 128, 203
  on human nature, 220, 235
  on natural selection, 18–19, 76, 80, 92
  on punctuated equilibria, 202

Gaastra, Wim, 185
Galápagos Islands, 36, 36n6, 100
Gale, Barry, 42, 44
Galton, Francis, 66, 78
genetic code, 195–97, 228, 231
genetic drift, 82, 84, 94
geographic speciation, 142–43, 200–201, 203; see allopatric speciation
Gerhart, John C., 216–18
Gilbert, Scott, 211
Gingerich, Owen, 236–37, 252, 259
Goldilock's enigma, 254, 258
Goldschmidt, Richard B., 27, 97, 130–31, 158, 161, 164, 171–72
  biographical sketch of, 139
  criticism of natural selection, 139–41
  Ernst Mayr's criticism of, 142–43
  on "hopeful monsters," 141, 145–46, 166
  and punctuated equilibria, 202–3

  ridicule of, 144–45, 183
  Sewall Wright's review of, 144
  Theodosius Dobzhansky's review of, 143
Gould, Fred, 114
Gould, John, 36
Gould, Stephen Jay, 104, 108, 153, 203, 212, 236
  on contingent evolution, 8, 52–53
  on non-overlapping magisteria, 11, 235
  on punctuated equilibria, 29, 199–202
  on Richard Goldschmidt, 139, 144–45
Gray, Asa, 42, 46, 56, 64
Greene, J. C., 189
Grose, Jonathan, 209

Haeckel, Ernst, 67–69
Haeckel's embryos, 67–69, 67n5
Haldane, J. B. S., 77, 84–85, 87–88, 90, 93, 140, 186n73, 213
  on orthogenesis, 137
  and population genetics, 15, 26, 83–84
  on teleology, 186
Hall, Barry, 174–75, 177–78, 180–81
Hamburger, Viktor, 145
Hamer, Dean, 12
Hamilton, William, 90
Haught, John, 7–9, 11
Heisenberg Uncertainty Principle, 250
Heisenberg, Werner, 244–45
Hennessey, Elizabeth, 36n6
Henslow, John, 34–35, 66
Himmelfarb, Gertrude, 32, 41, 213
HMS *Beagle*, 24, 35–36, 40, 66, 261
Ho, Mae-wan, 207–8
Hoekstra, H. E., 117n63
Holmes, Sherlock, 93
Hooker, Joseph Dalton, 42, 45
Hooper, Judith, 122, 125
"hopeful monster," 141–42, 144–45, 166, 183
Hoyle, Fred, 256–57

Humboldt, Alexander von, 34
Huxley, Julian, 16, 91–92, 96, 104, 109
Huxley, Thomas Henry, 16, 64, 91, 163, 183
  as Darwin's bulldog, 25, 47–49
  Ronald Fisher's use of, 88–89

industrial melanism, 6–7, 30, 96, 117–29
inheritance of acquired characters, 37, 43, 74, 177, 181, 204, 208, 210
Institute for Creation Research, 162
intelligent design (ID), xi–xii, 1, 30, 61, 115, 183, 199, 209, 211, 235, 243, 246, 259
  combinatorial inflation and, 113, 225–27
  criticisms of, 9, 19–21, 216, 230–31, 252, 261
  influence of, 229
  irreducible complexity and, 224
  Simon Conway Morris's criticism of, 54
  specified complexity and, 227–29
  summary of, 219, 221–23
  Thomas Nagel's view of, 23
irreducible complexity, 223–25, 230

Jablonka, Eva, 30, 199, 204–10, 215, 217
Jablonski, David, 151
Jacob, François, 29, 55, 191–93
Jensen, Derrick, 247
Johannsen, Wilhelm, 70
Johnson, Norman A., 115–17
Johnson, Phillip, xii
Jones, John, 222
Judson, Horace Freeland, 167–68
Jung, Carl, 167

Kadanoff, Leo, 21
Kaplan, Jonathan, 93
Keller, Evelyn Fox, 170
Kelly, Edward, 249

Kemp, T. S., 156–58
Kettlewell, Bernard, 6, 121, 123–24, 127–28
  criticisms of, 119–20
  experiments on peppered moths, 118
Kingsolver, J. G., 116–17
Kirschner, Marc W., 216–18
Kitzmiller v. Dover Area School District, 222
Knoll, Andrew H., 148
Kröger, Ronald H. H., 63
Kropotkin, Peter, 67
Kuhn, Thomas S., xiv, 94, 104, 120–21

Lamarck, Jean Baptiste (or Lamarckism), 37–38, 43, 71, 135, 211, 213, 217, 261
  August Weismann's criticism of, 75, 188
  and the Central Dogma of molecular biology, 166–67
  Denis Noble on, 210
  and directed mutation, 176–77
  and epigenetic inheritance, 30, 199, 204, 206, 208
  influence on Darwin, 36–37
  on the inheritance of acquired characters, 25
Lamb, Marion J., 30, 199, 204–10, 217
Lambert, David, 123
Larson, Edward, 85
Lenin, 84
Lenski, Richard, 176–78
Lessl, Thomas, 48
Levinton, Jeffrey, 148
Lewinton, Richard, 233
Linneaus, Carl, 133, 183
Luria, Salvador, 173–74
Lyell, Charles, 36, 39–41, 47n30, 64
  influence on Darwin, 37
  reaction to Darwin, 45–46, 56–57
  on uniformitarian geology, 35, 228
Lyons, Sherrie, 48

Lysenko, Trofim, 167

Mach, Ernst, 90
Mackenzie Presbyterian University, 230
MacPhee, Donald, 178–79
Majerus, Michael E. M., 119, 123–24, 126
Malthus, Thomas, 38–39
Marchant, Jo, 249
Martens, Koen, 207
Marx, Karl, 84
Matthews, Robert, 121
Matzke, Nicholas, 21
Mayr, Ernst, xiii, 4, 29, 45, 51, 100–101, 109, 128, 128n87, 139, 164, 201, 214
    advocacy of Darwinism, 1, 45, 92, 98, 100, 129
    on the Cambrian explosion, 149–50, 153–54
    criticisms of, 2, 3
    and the grand narrative of Darwinian triumph, 13–17
    on Haeckel's embryos, 68
    and the modern synthesis, 90, 95–96, 202
    on polyploidy, 160–61
    on population genetics, 86
    on punctuated equilibria, 203
    on Richard Goldschmidt, 142–43
    on Robert G. B. Reid, 213
    on scientizing biology, 85, 91, 105
    teleological statements of, 55, 111, 187–89, 192, 194, 196
    on Thomas Hunt Morgan, 74
    on the weaknesses of Darwinism, 42, 129
Mazur, Suzan, 215
McClintock, Barbara, xiii, 125, 181, 183, 216
    biographical sketch of, 169–70
    on cellular cognition, 172, 175
    on the Central Dogma of molecular biology, 28, 169, 172–73
    on transposable elements, 163, 171
Medawar, Peter, 102–3
Mendel, Gregor, 14, 25, 70–72, 74, 78
metaphysical materialism (or naturalism), 245–46, 248
methodological materialism (or naturalism), 245–46
Meyer, Stephen, xii, 23, 221, 228–29
Mill, John Stuart, 46, 49, 86
Miller, Craig, 123
Miller, Kenneth, 238–40
Miller, Stephen, 173–74, 179
Minnich, Scott, 221
Mittler, John, 178
Mivart, St. George Jackson, 53, 55, 59, 61, 72, 106, 138
    criticisms of Darwin, 49–51, 57
    on evolutionary convergence, 50–51, 55, 111
    on eye evolution, 57
modern synthesis (or modern evolutionary synthesis), 32, 55, 90, 104, 108–9, 121, 128, 164, 199, 242
    Bernhard Rensch and, 130–31
    challenges to, 29, 101, 139, 198, 203, 216
    and epigenetic inheritance, 207
    and the evolution of novelty, 211–16
    founding of, 16–17, 26–27, 94–96, 130
    and population genetics, 86–87
    Richard Goldschmidt and, 139, 142, 144–45
molecular clock dating, 149, 150–51, 153, 155
Monod, Jacques, 167
Montefiore, Hugh, 7
Morgan, Thomas Hunt, 15, 90, 97, 139, 142
    biographical sketch of, 73

criticisms of neo-Darwinism,
    76–77
  Ernst Mayr on, 74
  as a Mendelian, 25
Mount Rushmore, 228
Müller, Gerd, 211, 216
multiverse theory, 258–59

Nagel, Thomas, 23, 222, 247
National Science Foundation, 89
natural genetic engineering (or
    NGE), 28–29, 163, 182–85
natural selection, xii, 5, 10–11, 13,
    30, 66, 70–72, 95, 129, 169,
    180, 197, 220, 241
  Alfred Russel Wallace and, 39
  as all-powerful force, 17
  August Weismann and, 26,
    74–76
  Bernhard Rensch and, 131–32,
    135–39
  and the Cambrian explosion,
    148, 154
  and the Central Dogma of
    molecular biology, 28
  as characteristic of Darwinian
    evolution, 4, 220, 237
  and combinatorial inflation, 226
  and correlated progression,
    156–57
  as a Darwinian hypothesis, 10,
    17, 38
  and directed mutation, 173, 176,
    178, 181
  and the eclipse of Darwinism,
    66, 69
  and epigenetics, 204, 206, 208–9
  Ernst Haeckel and, 67–69
  in Ernst Mayr's definition of
    evolution, 3
  and the evolution of novelty,
    211–15, 218
  and eye evolution, 57–58, 61
  and field studies, 114–17
  George Gaylord Simpson and,
    104, 107–9
  George Ledyard Stebbins and,
    55–56
  Graham Bell and, 111–14
  and industrial melanism, 6–7,
    118, 120, 122–24, 126–28
  and intelligent design, 221, 224,
    229
  J. B. S. Haldane and, 83–84
  John Endler and, 109–10
  and the modern synthesis,
    96–97, 130, 164
  and natural genetic engineering,
    184–85
  and the *Origin of Species*, 13,
    24, 65
  and polyploidy, 159, 160–62
  and population genetics, 15, 26,
    77, 84, 86–88, 93–94
  Richard Goldschmidt and, 139
  Ronald Fisher and, 78–80
  Santiago Ramón y Cajal and, 64
  and the scientizing of biology,
    18–19, 90–93
  Sewall Wright and, 81–82
  St. George Jackson Mivart's
    criticism of, 49–51
  teleological analogies for, 29,
    187–95
  theistic evolution and, 238
  Theodosius Dobzhansky and,
    97–101, 244
  Thomas Hunt Morgan's criticism
    of, 73, 77
  Thomas Huxley's skepticism of,
    25, 47–48
  as an unconvincing hypothesis,
    27, 42–56, 64
neo-Darwinism, 74, 76, 176, 178,
    210–11, 215
Newman, Stuart, 141, 211, 215
Newton, Isaac, 253–54
Nilsson, Dan-Eric, 58–61
Noble, Denis, 209–11
non-overlapping magisteria (or
    NOMA), 11, 235–38, 240
Norenzayan, Ara, 12

On the Origin of Species, xiii, 1, 13,
    39–40, 42–44, 49n40, 56–57,
    64, 66, 75, 86, 97, 263
  Adam Sedgwick and, 49
  Asa Gray and, 46
  the Cambrian explosion and,
    146
  Charles Lyell and, 45, 56
  evaluation of, 24, 39–49, 65, 231
  Gertrude Himmelfarb and, 32
  John Haught and, 8
  John Stuart Mill and, 46
  St. George Jackson Mivart and,
    50
  Thomas Huxley and, 25, 47–48
Orr, H. Allen, 63
orthogenesis, 55, 83–84, 136–37,
    185
Overbaugh, Julie, 173–74, 179
Overton, Jacob McC., 115

Palevitz, Barry, 20
Paley, William, 219, 223
panpsychism, 82, 182, 248
Parisod, Christian, 163
Park, Lisa, 20
Pauling, Linus, 166
Peacocke, Arthur, 9
Pelger, Susanne, 58–60
Pengfei, Tian, 227
Pennisi, Elizabeth, 215
Pennock, Robert, 245–46
Penny, David, 184
peppered moth
  as icon of evolution, 6, 6n17, 128
  industrial melanism and, 117–
    19, 121–27, 129
Perlman, Robert, 218
Peterson, Michael, 10, 11, 13
phyletic gradualism, 200–203
Piatelli-Palmarini, Massimo, 215
Pigliucci, Massimo, 88, 93, 215–16
pleiotropy, 77, 135–36
polyploidy, 30, 130, 159, 164
  allopolyploidy, 160–61, 163
  autopolyploidy, 160
  definition of, 159–60
  Ernst Mayr on, 160–61

  as a non-Darwinian process, 83,
    162, 183
  Theodosius Dobzhansky on, 159
  in *Tragopogon miscellus*, 162–63
  transposable elements and, 163,
    172
population genetics, 22, 94, 129,
    203, 248
  assessment of, 84–88
  and directed mutation, 178, 185
  founding of, 15, 26, 77
  J. B. S. Haldane and, 83–84
  the modern synthesis and, 95
  Richard Goldschmidt and, 140
  Ronald Fisher and, 78–80
  in the scientizing of biology, 93
  Sewall Wright and, 81–83
  Theodosius Dobzhansky and,
    15, 26, 96, 98–99, 101, 128
Porritt, G. T., 126–27
Provine, William B., 13, 82, 86
  on evolution and ethics, 220,
    235
  on population genetics, 85
Ptolemy, 252
punctuated equilibria, 29, 199–203,
    212

quantum entanglement, 250–51
quantum evolution, 27, 107–8, 202
quantum indeterminacy, 240,
    250–51

Ramón y Cajal, Santiago, 64, 138
Reid, Robert G. B., 29, 61–62,
    212–14
Rensch, Bernhard, 16, 27, 96–97,
    109, 130, 140, 148, 164
  assessment of, 131–39
  biographical sketch of, 131
  on the necessity of multiple
    coordinated mutations,
    135–36
  on the relative age of taxonomic
    categories, 133–35, 146
Richards, O. W., 81, 81n28, 97
Robson, G. C., 81, 81n28, 97
Rockefeller Foundation, 89

Rose, Christopher, 214
Rue, Loyal, xi
Ruse, Michael, 10, 78

Saccheri, I. J., 127
Sagan, Carl, 233
Salisbury, Frank, 226–27
Salvini-Plawen, L. V., 194
Sanford, John, 80
Sargent, Theodore, 123–26
Schaab, Gloria, 9
Schindewolf, Otto, 131
Schlichting, Carl D., 218
Schlutter, Dolph, 161
Schwartz, Jeffrey, 199
Sedgwick, Adam, 49, 49n40
Sepkoski, John, 149
sequence capacity, 227
Shallit, Jeffrey, 21
Shapiro, James A., 2, 4, 13, 30, 185, 195–96, 216
  on adaptive immunity, 182–83
  on evolution as a mystery, 1, 231, 260
  on natural genetic engineering, 28, 181–84, 192
Shaw, George Bernard, 66
Simpson, George Gaylord, xiii, 135, 164
  on anthropomorphism and evolution, 194
  assessment of, 108–9, 129
  biographical sketch of, 103–4
  on evolutionary rates, 104–5
  on the fossil record, 106–7
  on the limits of materialist philosophy, 22, 262
  and the modern synthesis, 16, 96
  on quantum evolution, 27, 107–8, 202
  on the randomness of mutation, 105–6
Sirius Passet fauna, 147
Slatkin, Montgomery, 176–77
Smith, John Maynard, 90
Smocovitis, Vassiliki Betty, 90–91, 93

Soltis, Pam, 162–63
specified complexity, 223, 227–29
Spetner, Lee, 221
Spitzer, Jan, 230
Stearns, Stephen, 19, 76, 80
Stebbins, George Ledyard
  on evolutionary convergence in plants, 55–56
  and the modern synthesis, 16, 96
  on polyploidy, 159–60
  and teleological analogies for natural selection, 188
Stewart, Jimmy, 52
Stonehenge, 228
Symonds, Neville, 175–76

Teilhard de Chardin, Pierre, 27
  criticisms of, 102–3
  and panpsychism, 82, 182
  Theodosius Dobzhansky and, 102–3, 242
teleological Darwinism, 186–97
teleology (or teleological), 187–94, 196
  and analogies for natural selection, 29
  Asa Gray and, 46, 64
  Barbara McClintock and, 171–72
  Charles Lyell and, 64
  and convergent evolution, 54–55, 111, 195
  and directed mutation, 177, 181, 186
  and epigenetic inheritance, 206, 209
  and evolutionary progress, 213
  and facilitated variation, 218
  and natural genetic engineering, 184–85
  Pierre Teilhard de Chardin and, 82, 102
  and polyploidy, 163–64
  Theodosius Dobzhansky and, 27
theistic evolution, 103, 238, 240, 242, 244
Thomson, K. S., 145

Tragopogon miscellus, 162
transposable elements, 163, 171–72, 185–86, 207, 209
Tutt, J. W., 118, 124, 126–27

uniformitarianism (in geology), 35, 37, 45, 228–29
Unity of Science movement, 90

Valentine, James W., 148–49, 151–52
Vienna Circle, 90
vitalism, 37, 64, 131, 166, 168
Vopalensky, Pavel, 60
Vorzimmer, Peter, 44

Waddington, Conrad, 210
Walcott, Charles, 147
Wallace, Alfred Russel, 39, 42, 64, 78, 188
Wallace, Bruce, 206
Watson, James, xiii, 27, 90, 164–66, 168
wave-particle duality, 249
Wedgwood, Josiah, 35
Weismann, August, 79
    criticism of Lamarck, 25–26, 74, 204
    as founder of neo-Darwinism, 74
    natural selection as default assumption, 75–76
    on the selective value of the initial stages of variation, 75, 101
    and teleological analogies for natural selection, 29, 188

Wells, Jonathan, 67, 221
Whitfield, John, 215
Whittington, Harry, 52
Wilberforce, Samuel, 47
Wilkins, Adam, 184–85
Wilkins, John S., 9
Wilkins, Maurice, 166
Wilson, E. O., xiii, 11, 220–21, 237
Wistar Institute, 113, 224, 226
Woese, C. R., 195–97
Wolf, Stewart, 21
Woollacott, Marjorie Hines, 249
Wray, Gregory A., 68
Wright, Sewall, 27, 83, 87–88, 90, 93, 98, 140
    differences with Ronald Fisher and J. B. S. Haldane, 84–85
    on effective population size, 81, 99
    feud with Barbara McClintock, 171
    feud with Ernst Mayr, 86
    as founder of population genetics, 15, 26, 77
    on genetic drift, 82
    and panpsychism, 82, 182, 248
    on Richard Goldschmidt, 144
Wrightsman, Bruce, 252–53

Yanofsky, Noson, 249, 260

Zygon: Journal of Religion and Science, 9–10

www.ingramcontent.com/pod-product-compliance
Lightning Source LLC
Chambersburg PA
CBHW021652230426
43668CB00008B/605